THIS BOOK FEATURES MULTIMEDIA CONTENT BEYOND THE PRINTED PAGE

Throughout this book are **special icons** you can use to activate and discover additional content on your smartphone, mobile, or tablet device.

How Does It Work?

1. Visit www.harpercollinsunbound.com to **download the free app** for your iOS or Android device.

2. When you see **this icon** FILM on pages throughout the book, open the app on your device and scan the page.

3. The app will do the rest, bringing multimedia and interactive content that relates to the page you're reading onto your device.

For the salmon of Devil's Gulch,

so they may run again for my daughter's children . . .

and those who come after.

CONTENTS

OUR AGREEMENT

I need to ask you a favor. After reading this book, please give it away. You'll know who it's for: Your mother. Your neighbor. Perhaps the person sitting across from you at work. This book desperately needs you. It needs you because these other people—the real audience for this book—may never buy it. They can't fathom the danger of living within an industrialized food system. For this reason alone, they need your help.

The intent of this book is simple. Seduce people with quirky collages and folksy handwritten notes that quietly introduce the tools to fix our crappy food system. The more people read, the more they'll remember. Ideas can be powerful like that.

So, the challenge remains. How do we get this book into other people's hands?

That's where you come in. Now that you've read these words, finish the rest of the book then tear out this page. You'll know what to do next. Let future readers remain oblivious to our silent pact. That's how movements work. Our identities remain a mystery, but our convictions are strong enough to topple empires.

A ROAD TO DAMASCUS MOMENT

One of my clearest recurring childhood memories is that of spawning salmon. They ran up Devil's Gulch, in Marin County, California, throwing themselves against the current, willing themselves over stones and flotsam-packed barriers of leaves and debris, onward past the limits of exhaustion, only to pause in deep pools, hidden beneath the oaks and towering redwoods lining the creek's steep banks, recharging in anticipation of yet another upstream surge. These salmon were here to spawn, lay eggs, then die. **This struggle of animal against nature was a yearly childhood event, one that connected me, even as a young boy, to the mysterious natural world around me.**

Decades later, I was justifiably excited to share this experience with my own daughter. When February came, my wife and I bundled her in a coat, scarf, and mud boots, and hiked the trail leading up from Lagunitas Creek to Devil's Gulch.

The salmon weren't there.

As I would later learn, this spawning ground had collapsed years before. Theories abounded. Overfishing. A golf course built beside a nearby stream. Runoff from a hillside housing project. No one knew for sure.

I never properly cataloged that disheartening winter afternoon until two years later, when I stood in a Georgia hay field talking about grass-fed cattle and the vagaries of summer rain with Bill Hodge, an earnest, calmly focused cattleman who ultimately turned our conversation—as often happens in these parts—to the Bible. In the years after Jesus was crucified in Jerusalem, Saul of Tarsus swore to wipe out the newly formed Christian church. He set off for Damascus armed with a letter from a high priest authorizing the arrest of any new followers. After a blinding light struck down Saul and his companions in mid-journey, a voice spoke, one only Saul could hear. Its message triggered a series of events that ultimately led to his religious conversion. **It was the first "Road to Damascus" moment. Hodge's own came ten years ago.**

Bill says "I'm the product of three different land grant institutions. All my formal training was based on the 'Industrial Ag Model'. Part of my job required me to educate other producers. While traveling [h]ome late one night in the mid 90's, after [giv]ing a presentation to beef producers on the [...] of the mainstream beef system, the futility [... I] was doing suddenly struck me:

[... n]ot put on this earth to be transported long distances, [...] yards and stuffed with concentrate feeds. They were [...]. That realization was my ...

DAMASCUS MOMENT

[...] oneself, denoting a change in attitude, perspective or belief

Bill Hodge
Georgia Cattleman

turning me to the tune of quite a sum of money. From that point forward I began researching more sustainable beef production (the

I'd been "stockering" 700 head at the time (putting weight on the cattle as efficiently as possible) when the bottom fell out of the market

[...]TIMIZED BY THE GOVERNMENT'S 'DAIRY BUYOUT PROGRAM' (PAYING PRODUCERS TO GET OUT OF THE BUSINESS BY REDUCING THEIR HERDS).

Back in the 1980s, Hodge was victimized by the government's "dairy buyout" program, which paid dairy producers to get out of the business by reducing their herds. He accepted their offer and began stockering his herd—putting weight on the cattle as quickly and economically as possible—when the bottom suddenly fell out of the market. He lost nearly seven hundred head. An industry he'd never questioned had failed him.

He thought about doing something else, but his love for meat animal agriculture was simply too great, so he ended up working in producer education for the local land grant institution. Ironically, part of his responsibilities included teaching cattlemen how to work within the "big corporate meat processor" model.

"I was traveling home late one night, after making a presentation to beef producers on the merits of the mainstream beef system," Hodge recounted, "and that's when the futility of what I was doing suddenly struck me. Cattle were not put on this earth to be transported long distances, confined in feed yards, and stuffed with concentrate feeds. They were put here to graze. That was my 'Road to Damascus' moment."

What happened to the Devil's Gulch salmon of my childhood? How many other people, folks just like me, have their own childhood stories of close encounters with a natural world that has since vanished?

We are living within the Great Change—it's happening at this moment—but the shift isn't sudden. It won't startle us into action. This is a gradual change. A creeping change, imperceptible from one day to the next. It's industrial, technological, and dehumanizing. We can't see the Great Change because it lacks immediacy; we tend to forget it's happening. The witnesses to these "before and afters" in the natural world, the people who still remember the "Way Things Used to Be," are disappearing, dying off. The amazing salmon runs of my childhood are gone from Devil's Gulch. That winter morning, as I attempted, without success, to show my five-year-old daughter precisely what she *wasn't* seeing in that creek, I had my own "Road to Damascus" moment.

I want to bring those salmon back. I want my daughter to share these memories, to understand the complexity of things, to one day restore the natural order to Devil's Gulch. To do that she'll need to understand the meaning and function of all those moving parts, the nameless pieces powering this modern industrial apparatus we've created.

We've all had Devil's Gulch moments. We've all felt powerless when confronted by systems composed of parts we depend on yet don't understand.

This is a book about first understanding, then dismantling, our industrial food system.

Can "the way it is" be reexamined, its offending parts held up to the light, observed, and discarded?

Our food system is opaque, composed of nameless objects. Deciphering their meanings will require mastering a new language, which can be learned one word at a time.

We live in a world of dwindling natural resources, but the principle of SUSTAINABILITY offers us a road map for managing what we have left. Yet as we attempt to put our world back in balance, we've seen the term sustainability grossly misused, its meaning devalued, hijacked, turned into hollow-sounding marketing jingles.

The stakes are high. If people don't understand the meaning (and implication) of terms like **FOOD MILES, CARBON FOOTPRINT, CSA, ORGANIC, FOOD SECURITY, FOOD DESERT, GMO, GRASS-FED, DIRECT TRADE**, or even **PASTURE-RAISED**, how can they live more sustainably?

To help "take back" the meanings of these important ideas, I set out to document the work of two hundred thought leaders, architects of a new vocabulary reflecting the most promising solutions for creating a vital and sustainable food system in this country. I asked each of them for a single term that defined the essence of their work, because words are truly the building blocks for new ideas. I feel like I've only captured a fragment of this veritable reimagining not only of American agriculture, but of our country's consumer culture, one defined by excess and disposability—but now one showing signs of responsibility, of resilience, and even of hope.

While my photographic subjects are varied—some folks are famous, others less so—they show that this reinvention of our food system has no central figurehead or single source for inspiration. It's happening everywhere, and it's happening now.

PASTURE RAISED VS. CAGE FREE

Alexis prefers to say that her chickens are "pastured" instead of "free range" because there is imprecision used to describe how chickens are raised. The term "cage free", for example, could be used to mean poultry raised indoors with little or no access to the outdoors.

April 23, 2008
Soulfood Farm
Vacaville, CA

The chickens eat grass and bugs and chicken food (and predators like red tail hawks, black birds and coyotes eat them)

1800 eggs per day

Alexis treats her chickens with love and a real respect for the gifts they give with their delicious eggs and meat. They are raised humanely and given all the room they need outside to be what they really are, and in return give us the best of what they are.

There are seven types of laying hens including rhode island reds, barrel rocks and delawares (the meat birds are colored rangers)

THE POWER OF WORDS

Go ahead. Tell me words aren't powerful. Or that words, by themselves, won't change anything. Tell me no one reads anymore. Tell me we're impatient, visually literate but crippled by short attention spans. Tell me you can't actually transform the world because the power elite have jury-rigged the system, that they control the levers and direct the machinery. Tell me you're powerless. That it's all too complicated. That nothing matters. Tell me that you have neither time nor money. That you're tired. All the time. That you don't have enough energy to do the things you want. Tell me our leaders don't listen anymore, our government is accountable to no one, and our Congress is gridlocked. Tell me you've stopped watching the evening news because it bears no resemblance to your daily life. Tell me you no longer remember the "Good Ol' Days," the last time you smelled a flower, walked barefoot on dew-kissed grass, picked apples from a tree, or felt the warming glow of sunlight on your face.

Tell me you simply don't care, because why should you? And I'll tell you the story of an egg.

Eggs come from the supermarket. You buy an egg and you eat it. One day an egg carton offers the words **CAGE-FREE**. You're puzzled, maybe because you have only the vaguest of notions about where eggs come from. Chickens in cages? Really? Cage-free must obviously mean a better life for that poor fluffy cartoon-faced chicken on the carton. And the egg? It might taste better, too. Other people learn about this new term and start buying cage-free eggs. The result? An entire industry—the retail marketplace being both ruthless and pragmatic—is forced to change its practices. It's astonishing. The Food and Drug Administration (FDA) can take no credit. It's all due to the work of time-starved, money-conscious, "powerless" consumers who didn't have time to watch the evening news. **They learned two simple words, "cage-free," paid an extra dollar at the supermarket, and forced an entire industry to reexamine the way it did business.**

By learning what a few words mean you can help reinvent our crappy food system. It's a battle won slowly, product by product, supermarket aisle by supermarket aisle. It starts with cage-free, and progresses to **FREE-RANGE** and **PASTURE-RAISED**. Each term represents the movement of a philosophical line in the dirt, a deeper commitment to principles, and a greater attempt to achieve transparency both in the farmyard and in the marketplace. Consumers want to know what's in their food and they want it printed right on the label.

If someone is willing to pay a few extra dollars for a dozen more ethically produced eggs, there is hope. That egg is the gateway to a larger conversation. **Consumers' willingness to align their values with those of the product they're buying—even when it's something as cheap as a dozen eggs—means the industrial food system is a house of cards.** It proves that when presented with clear, compelling explanations about what they're eating, consumers will make purchases that reflect their ever-expanding **FOOD LITERACY**. When consumers shop and eat according to their values, the food industry is forced to adapt.

CAGE-FREE: Chickens that are not kept in cages but still confined to a barn with limited or no access to outside.

FREE-RANGE: In the United States, USDA regulations apply only to poultry and indicate solely that the animal has been allowed access to the outside. These regulations do not specify the quality or size of the outside range nor the duration of time the animal must be allowed access to this space.

PASTURE-RAISED: Animals that have been raised on pasture with access to shelter. This term is being used by farmers who wish to distinguish themselves from the industrialized "free-range" term.

rBST FREE

rBST = recombinant bovine somatotropin   rBGH = recombinant bovine growth hormone

Both rBST and rBGH are terms used to describe synthetic growth hormones which increase milk production in dairy cows.

The USA is the lone remaining country in the developed world which still permits the sale and use of rBST. A number of well-documented consequences of its use (ranging from increases in clinical mastitis to infertility in dairy cows exposed to the hormone) has led a host of nations to ban its use: Canada, New Zealand, Australia, Japan and all 27 member countries of the European Community.

The court also cited 3 reasons milk produced by rBST-treated cows is different: increased levels of IGF-1 hormone; period during each lactation with lower nutritional quality in milk; and increased somatic cell counts (i.e. more pus in the milk).

IN 2010, THE 6TH CIRCUIT COURT RULED THAT IT'S LEGAL FOR MILK PRODUC

necessary to serve the state's interest in preventing consumer

Furthermore, they said Ohio's ban on dairy processors' hormone-free claims violated their 1st Amendment rights and was "more extensive than

somewhere in America
18 July 2011

TO LABEL THEIR MILK AS "rBST FREE."

The industry may file lawsuits. They can paralyze the FDA and slow the inevitable by obscuring facts and even confusing consumers. Or in the case of milk they may remind you—repeatedly by legal decree—that the "FDA maintains that no significant difference has been shown between milk derived from **rBST**-treated and non–rBST-treated cows." Despite all that, the story already has its ending. The public has already learned about rBST too, a synthetic growth hormone that increases a cow's milk production but can cause health problems in the animal. They read about it on the Internet, notice it on a milk carton, or overhear a conversation about it at a farmers' market. In short, they become educated, up their food literacy, and reach their own conclusions about which milk to buy.

Retailers across the United States have taken notice. Want proof? Scan the refrigerated case at Costco, Safeway, Walmart, Kroger, Publix, and a hundred other supermarkets: **You won't find any milk here made using rBST. Consumers learned a simple term, paid a little extra at the cash register, and changed an entire industry. Again. All because of the power of words.**

And it may happen yet again. **GMOs** are genetically modified organisms. They're in most of what we eat here in the United States, and that makes us unique. We are one of the few industrial nations in the world that still allows companies to sell food loaded with GMOs without saying so on the label. It may take generations to fully grasp their environmental and health impacts. Until that time, many consumers simply ask for transparency. They want to know what's in their food, and they want it in writing. Again, when consumers increase their food literacy and eat according to their values, the food industry changes.

IBM once ruled the computer universe, and people believed its hegemony would last forever. Then Microsoft appeared. Every empire—be it a company or a nation—may seem preordained to last a thousand years, but change comes to everything. Companies that adapt survive. Those that don't lie buried and forgotten, destined to be dug up a thousand years from now and studied for clues hinting at the haphazard series of events that led to their demise.

LOCAL
Consumers who support local food producers strengthen local economies

CORNELIUS

WALTER

GUS

MARIE

BARBARA

The Village Grocer

TOWN

LOCAL FOOD PRODUCERS

LOCAL FOODSHED
A region where food is grown and consumed:
1. land where food grows
2. local distribution system
3. markets where it is sold
4. people who consume it

POLYCULTURE VS. MONOCULTURE
Farmers who grow a diversity of fruit and vegetable crops in one place (polyculture), instead of a single crop like corn or soybeans (monoculture), provide more local food choices for their communities.

EATING IN SEASON
Consumers who eat local foods are eating with the seasons.

LOCAL FOOD HUB
A centrally-located facility that provides for the aggregation, storage, processing, distribution, and marketing of local food products.

INSIDE FARMVILLE 2
SOMEWHERE ON THE INTERNET
27 NOVEMBER 2012

Marginal text (top): game also includes many exotic livestock ranging from Chianina cows to Polish Silver Laced chickens. Players earn points.

Marginal text (left): both alone and with other players by harvesting crops and creating value-added products like Elderberry Tea and Apple Sauce.

Marginal text (right): The game features over fifty types of fruits and vegetables, plus another fifty types of trees, with new varieties added each week. The

Marginal text (bottom): FARMVILLE2 IS AN ONLINE GAME THAT REWARDS PLAYERS FOR GROWING A VARIETY OF CROPS AND PRODUCING GOODS FOR THEIR COMMUNITY.

In early November, long after its last leaves have fallen, the persimmon tree down the road from our farm begins to fruit. Its orange globes hang from naked branches, looking like overly enthusiastic Christmas ornaments come a month early.

We pay our neighbor a dollar for each one we pick, and she always tells us it's a good deal. Maybe so. I just like stopping every few days, surveying the fruit, and choosing which ones to pick on the way home. The firm ones get grilled and added to salad. The ones that get really ripe, with their insides turned to jelly, are eaten with a spoon. For our family, persimmons mean a tree on Middle Two Rock Road.

We can't have that same intimate relationship with everything we eat, because our food mostly comes from somewhere else, and it's grown and prepared by people we'll never meet. Most of us will never pull a carrot from the ground, milk a cow, slaughter a pig, or gather eggs from our own hens. Those days of rugged self-sufficiency are gone and aren't likely to return. Yet people are increasingly aware that their hyper-accelerated, super-improved lives are missing something. They're rethinking not only what they eat, but where it comes from. This crusade has a name: the **LOCAL FOOD MOVEMENT**.

FOOD MILES

One evening I find myself dining in Harrisonburg, a well-heeled Virginia town an hour west of Washington, D.C. The dinner guests include a few local winemakers and a food policy wonk who patrols Capitol Hill. He talks about food-related legislative challenges—GMO labeling issues, oversight of factory farms, school lunch vouchers—then veers into a colorfully detailed anecdote about the lengths discerning Washington shoppers must go to find good food inside the Beltway. He uses the term *food miles* to describe the great distances he drives each week just to buy groceries. The man is intriguing and well-spoken, but his definition of food miles is backward. **FOOD MILES** are not determined by how far *you* drive to get your food; instead, they are the energy and effort required to bring everything you eat to your door. The milk. The bread. The mustard. The oatmeal. The sugar. The potatoes. All of it comes from somewhere; that distance is called food miles.

If a Washington, D.C.–based food policy advocate doesn't understand the definition of food miles, what hope do we have for fixing our food system? After all, we can't expect people to live more sustainably if they don't know the most basic principles that define the conversation.

We're the first generation forced to confront a fundamental truth: We live in a world of finite resources. Changing how we consume these precious resources—safeguarding them for future generations—will require not only changing our behavior, but even learning a new language. We need new words

to explain the predicament we're in and what we're going to do about it. Scientists talk about **PEAK OIL** and **PEAK WATER**, conditions that signal a point where our consumption will outstrip the availability of remaining resources. Economic collapse is their dire prediction. **They could be right, but scare tactics won't bring about a shift in consciousness. People run from bad news, not toward it. The answer is to build consensus on a foundation of innovative ideas and solutions. Find out what works, then let solutions spread.**

It all begins with words. By learning the words of this new language—the lexicon—you can start the conversation, even embrace ideas that had previously seemed foreign or irrelevant to your daily life. If you start by learning what the term *food miles* means, for example, the transformation begins. These words are building blocks, which is ironic considering that once you understand the principle of food miles, you inevitably see flaws in the concept it represents. It's attention-grabbing and visual, but ultimately imprecise. You may begin to prioritize local purchases over those from far away, which is good . . . for starters, but the principle of food miles is merely a point of departure. It forces you to think about your connection to the soil and to local economies. It's a lesson more useful in theory—as a teaching tool—than in practice.

Take the tale of two satsuma oranges. Those at your local farmers' market come from twenty miles down the road, whereas your local supermarket has them delivered by a tractor-trailer from five hundred miles

"**PEAK OIL** is the moment in time when the world rate of crude oil production reaches a maximum and begins to decline. The peak will probably be driven by a combination of geological, economic, and political factors. Since oil powers nearly all transportation and transport is key to trade, if the peak happens soon (before substitutes are found and deployed) the result will almost certainly be sharp economic decline."
—Richard Heinberg, Post Carbon Institute

"**PEAK WATER** is the idea that there are ultimate limits to our ability to take and use water from natural systems. For rivers, streams, and many aquifers, humanity is reaching (or passing) those limits now."
—Peter Gleick, Pacific Institute

away. That delivery brings enough satsumas for a hundred stores. Simple mathematics tells you that less fuel is used to deliver supermarket satsumas. At least you've met the farmer at your local farmers' market. Plus, he's organic. Except the satsumas at the supermarket are certified organic, too. Right, but those satsumas aren't *local*. You don't know anything about them. So is it a question of relationships and trust, or a question of scale or the optimum allocation of resources? The satsuma you buy often explains the personal values that drive your purchasing decisions. Do you want food that's local, organic, or just cheap?

One important consideration when buying locally grown produce is this: Fresh produce is high in phytonutrients, naturally occurring chemicals that provide a variety of health benefits. The longer produce sits in a truck, the more phytonutrients it loses. For example, spinach loses more than 50 percent of its vitamin C one day after harvest. So if we want healthful food, doesn't freshness trump everything?

These are complex questions, but there's beauty inherent in this complexity. The more questions this new knowledge forces us to ask, the more we start understanding that every purchase we make is a vote to support one system over another. Do you believe in a global food supply chain or do you want a more local, fresh alternative that supports our neighbors and their sustainable farming practices, keeps our waterways clean, and our soil healthy?

It's not simply about choosing one satsuma over another. It's about knowing the tale each satsuma represents.

Until a few years ago I wouldn't say I was a particularly enlightened consumer. I bought things because they tasted good, because they were on sale, or because I knew what to do with them in the kitchen. I understood a little bit about nutrition. I knew fruits and vegetables were boring but good for me. Soda was full of high-fructose corn syrup, which sounded ominous though its supposed dangers eluded me. I knew potato chips contained an artery-clogging dose of partially hydrogenated oil and that eating steak every day would give me a heart attack. I knew white bread was packed with empty calories and carbohydrates designed to make me fat. Even with that meager food knowledge, I still didn't change what I ate, and I certainly never stopped to think about who made my food. How could I? I spent most of my life living in urban areas. My daily life featured an endless panorama of glass and steel high-rises, billboards, parking structures, stoplights, crosswalks, cars and more cars. I never thought about where my food came from or who made it. And I certainly didn't know any farmers—I couldn't even find a farm on the map. I was utterly disconnected from who made my food and how it ended up in my grocery store. The fact was, I simply didn't care.

When I'd meet one of those Know-It-All Do-Good Lifestyle Fascists, with their clever slogans designed to shame me into turning off lights when I left the room or recycling plastic bags or turning the water off when brushing my teeth or only eating organic or not wearing leather belts or not buying slave-trade coffee, I'd sigh, smile knowingly, then calmly dismiss them as elitist. They were simply out of touch. Only wealthy people have the luxury of recasting the world to fit their own values, I'd say to myself. Food is just something you eat.

What changed for me? I learned to cook and, as a result, I became aware that a dish is composed of many individual ingredients, each with its own story, like that of a cage-free egg or the tale of two satsuma oranges.

The term *food miles*, even with its contradictions, opens a gateway. It establishes that food isn't merely a commodity. A person grew it. Feats of dedication and logistical handiwork allowed it to travel great distances. And *you* factor into this equation too. After all, you are the beneficiary of all this labor and largesse. Those carefully stacked pyramids of fruit in the produce aisle have been placed there just for you. If your allegiance is elsewhere, if the miles your food travels matter and you'd prefer instead to buy your food from a local farmer, the entire distribution apparatus will change.

TERROIR

I pull into Staunton, Virginia, late one evening. The town is a small collection of low brick buildings dating back to the 1800s. You might call it charming or quaint or bucolic, but I fail to make these assessments. I'm tired and irritable; all I can think about is food. I've driven seven hours without a stop—having resisted the sirenlike lure of fast-food beacons marking my journey down highways 66 and 81—and now I'm starving. At this hour, everything is closed. I end up at the Stonewall Jackson Hotel, where the bellman tells me about Zynodoa. The restaurant might still be open.

I find it just around the corner; it's small, a single room with maybe a dozen tables, and it's closed. But the front door is ajar, so I push it open; entering I hear music from the kitchen. Then a waitress carrying a stack of napkins emerges from a hallway. She tells me they're closed for the night, but I'm persistent. I tell her I've driven for hours and will eat anything left in the kitchen. She sighs, directs me to a stool at the bar, and fetches me a plate. It's then that I notice the far wall. It's dominated by a chalkboard listing all the local produce and meat on offer that night; it's impressive. **This idea of locally grown food is real and it's a movement; there's no one single spokesperson or geographic center. It's happening everywhere, and it's happening now, even in towns you never knew were on the map.**

A man pokes his head out the kitchen door—the chef, perhaps?—then disappears. A few minutes later he comes over with two plates. The chef—his name is James Harris—sits down beside me and we dig into grilled chicken breasts. After noticing that the chicken comes from Polyface Farm, I tell him I'm spending the next day with its proprietor, Joel Salatin (see pp. 172–177). Harris talks about Salatin's farm, the inspiration he brings to this rural community, then about the other local food producers who supply Zynodoa.

"It must make you proud to see all their names up on the wall," I surmise aloud. "To have so many local food producers you know by name and who grow so many things."

Harris just stares at me. "Actually, it's kind of sad," he counters. "What does it mean about how we live that I actually have to tell you that the food came from here? That I need to write it on the wall, tell you everyone's name, and make it out to be something special. If you go to a place, you should expect to eat what grows there. Am I happy to list everyone on that wall? Honestly, I'd be happier if I didn't. If you just came here and knew the food you ate was from the Shenandoah Valley—that it told you something about where you are, and in what season, and how we do things here—we wouldn't need that wall. The food would just speak for itself."

This notion that food has specific qualities defined by a sense of place is called **TERROIR**. It's a French word, one often used to describe not *how* wine tastes but from *where* it tastes, and not from a winemaking region but from a single vineyard, even a single lot planted on a single hill. It's that precise.

Scientists might say terroir is determined by unique mineral combinations in the soil or an area's microclimate, which is akin to a climatic signature. A farmer's growing practices may also play a role, which I discovered when visiting Judith Redmond at Full Belly Farm in California's Capay Valley.

This concept is not limited to the soil. In trying to describe what made his West Marin oysters so special, Kevin Lunny grasped at a number of adjectives before settling on the term **AGUOIR**. He coined the word by combining the principle of *terroir* with *agua*. Never mind that his term to describe a distinctly Californian estero steeped in aquaculture traditions was comprised of two incongruous words—one French and the other Spanish—bolted together; for Lunny, the estero's aguoir is unique to America's West Coast, being the only inlet totally protected from pollution by urban storm sewers and farmland runoff.

I remember that oyster farm from my own childhood. Years ago it was run by the Johnson family, who often served oysters the local way, set atop a barbecue grill, steaming until they popped open on their own, a tradition updated in recent years to include a dab of Tapatío sauce.

Full Belly Farms
Capay Valley (Guinda) CA
27 January 2012

Each farm is unique so the combined effects of
terroir take time to be revealed. Some parts
of a farm are colder in spring. Some are weedier.
Others... sandier. A farmer must understand
the land, meet new challenges as they arise
each year and develop a farming approach
that builds resilience.

navel oranges

"terre"¹ = land
1 (a French term meaning all food
expresses a sense of place)

TERROIR

flavor + character of produce = climate + geography + farming practice

At Full Belly Farm, organic farming and soil building practices combine with
the terroir of the Capay Valley to build deep flavor in their fruits and vegetables.

AGUOIR*

formula for term: $terroir^1 - terre^2 + agua^3 = aguoir^*$

$terroir^1$ = a French term expressing the idea that food tastes of its place

$terre^2$ = soil (French) $agua^3$ = water (Spanish)

Kevin tells me that the oysters here taste differently depending on where they are raised in the estero. He calls this "aguoir."*

"singles" are raised in "grow out" bags set out in the estero's intertidal zone. As the oysters grow they are placed in increasingly larger bags until harvesting, at 16-24 months.

While the oysters are still small they are attached to shells that are then threaded onto cables. These "cluster" oysters will hang from racks for 18 to 24 months until harvesting.

TEODORO

KEVIN

PIPPIN

Oysters have long been a fixture of West Marin life. Most weekdays, when the tourists aren't around, locals still gather a few miles up the road from Point Reyes Station at the Marshall Store for a dozen Kumamotos and a beer.

Europe has long-held traditions that bind food to geographic regions. Champagne comes from Champagne; Armagnac from Armagnac. In addition to wine, cheeses and some types of meat are specifically defined by the regions—or appellations—where they are produced. Italy's Reggio Emilia region provides us with both prosciutto di Parma (a Parma ham) and Parmigiano Reggiano (a Parma cheese). Both are geographically protected foodstuffs, and their product names are protected from misuse and imitation by international courts of law.

In the United States these ideas are less formal. A stamp may certify that Vidalia onions come from an area in Georgia, but such geographic identifiers are largely reserved for marketing-speak. We have Idaho potatoes, Washington apples, California cheese, Florida oranges, and corn from, well, everywhere.

Still, where food comes from matters. Indigenous food traditions lead to seasonal events that further deepen a culture's connection to food. On any given weekend, *sagras*, or food festivals, take place across Europe. The continent is both defined by and celebrated because of the diversity of its food. It's the same in most other parts of the world. Today, food still marks the seasons and reaffirms a culture's identity, but as farming communities adapt to monoculture crop production and as the population exodus from rural to urban communities accelerates, these cultural traditions slowly erode and diminish in relevance until only faint recollections of them remain. One day they too will vanish, leaving us with endless shelves filled with the same gaudily packaged vacuum-packed treats stacked in cavernous fluorescent-lit food warehouses that line the closest freeway off-ramp. **These big-box supermarkets don't sell terroir. How could they when these companies themselves have no place to call home?** They're open twenty-four hours a day, seven days a week, and are ready to sell you any food product you could possibly imagine. And there's nothing special about any of them.

KEVIN AND NANCY LUNNY'S BARBECUED DRAKES BAY OYSTERS
Serves 4

INGREDIENTS:
¼ pound (1 stick) salted butter
4 cloves garlic, minced
½ teaspoon freshly squeezed lemon juice
¼ teaspoon minced parsley
16 medium oysters, preferably Drakes Bay
Cocktail sauce

DIRECTIONS:
1. Preheat an outdoor grill to high.
2. Under running water, clean the oyster shells using a brush.
3. Shuck the oysters, removing the top (flat) shells while leaving the oysters and their liquor in the bottom shells; discard the top shells.
4. In a small saucepan set over medium-low heat on the stovetop, melt the butter, and add the garlic, lemon juice, and parsley.
5. Place the oysters, shell side down, on the preheated grill.
6. Spoon the melted butter mixture over the oysters and cook until the butter just begins to brown around the edges of the shells.
7. Place a small dollop of your favorite cocktail sauce on top of the oysters. Serve immediately and enjoy with your beverage of choice (a good Sauvignon Blanc for Nancy and a Lagunitas IPA for Kevin).

A NEW MARSHALL PLAN, or
ECONOMIES OF COMMUNITY

By the end of World War II, cities, bridges, power stations, railways—the entire infrastructure of Nazi Germany—had been destroyed by Allied bombers. Then the United States helped rebuild everything. This was called the Marshall Plan.

Back in America, the postwar era created an economic boom. A new Interstate Highway System was constructed, allowing businesses to expand across the nation. It was an awesome undertaking, the greatest public works project since the building of the Egyptian pyramids. This latticework of roadways opened new markets. Small businesses got bigger, and big businesses became multinational.

David Bauer, a baker in Asheville, North Carolina, once showed me a one-hundred-year-old map of the East Coast. Each dot on the map—and there were hundreds of them—represented a grain mill. At one time there were two or three grain mills in every county, sometimes more. Wheat was primarily grown and milled locally.

By centralizing production, companies in Middle American towns could bake bread and truck it across an entire nation, enabling them to sell their product at prices cheaper than a local baker. Cheaper bread, even by just a few cents a loaf, eventually meant no local bakers. That meant no local wheat, which meant no local mill, which in turn meant that the knowledge to grow and mill that wheat was lost as well.

Since the end of World War II we've witnessed the consolidation of nearly every aspect of our food system. Across the country, the vital local infrastructures that once supported and fed communities, that took decades to build, have been dismantled. Towns have watched their slaughterhouses, supermarkets, butchers, dairies, and bakeries simply disappear.

The U.S. government has enacted laws to protect consumers from monopolistic practices, but industries still consolidate. Once they do, they rarely come unraveled. Phone companies and rail trusts may have been broken up, but the food industry is more complex. The interlocking relationships that define this industry are harder to discern. **Can an industry that consolidates be pulled apart?**

I put that question to Rich Pirog, senior associate director of the Center for Regional Food Systems at Michigan State University. He thinks industries can definitely shift from decentralized to centralized, then back again.

"Even though it's not food, one of the best examples would be the music industry," Pirog says. "There was a period of time that it was very decentralized. You had orchestras and traveling bands. Music was local. Then we got to the point where five companies controlled the music industry in this country. The advent of Napster and other technologies moved the music industry to a much more decentralized approach."

Napster recast the Internet as a disrupter capable of toppling monolithic markets, but with mixed results. Although it unleashed piracy of intellectual property on an unprecedented scale, it also proved that technology placed directly in consumers' hands could effectively eliminate the middleman, open up distribution channels, and level the playing field for new entrants. So is Pirog banking on the wisdom of crowds? Is he saying that consumers, when given access to the right tools, can collectively balance free markets?

Eric Holt-Giménez, executive director of Food First/Institute for Food and Development Policy, frames

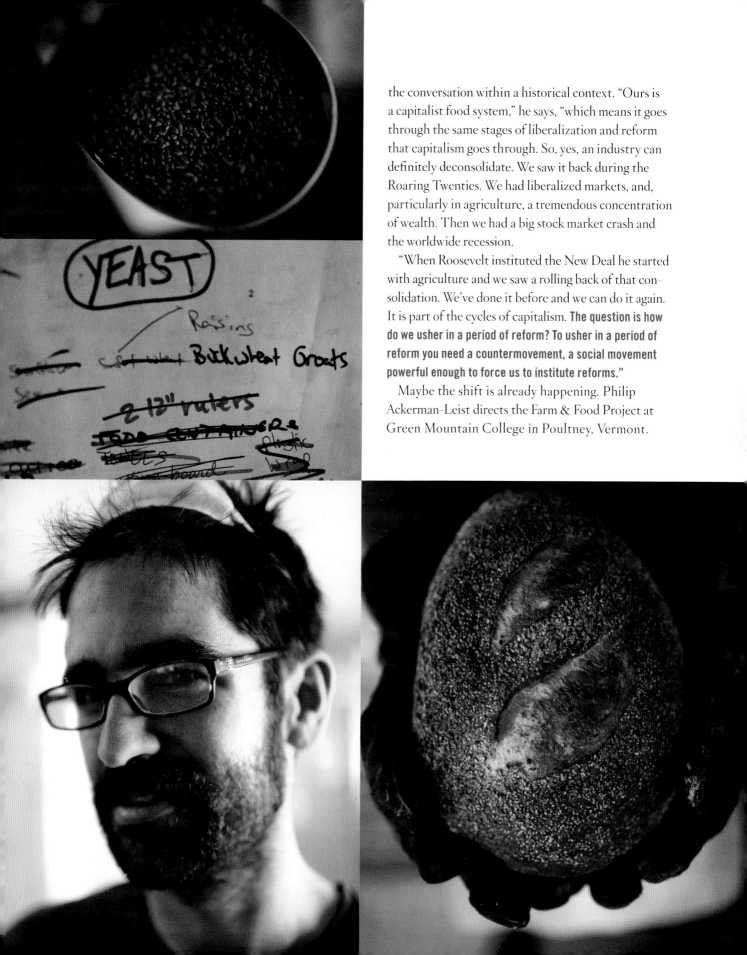

the conversation within a historical context. "Ours is a capitalist food system," he says, "which means it goes through the same stages of liberalization and reform that capitalism goes through. So, yes, an industry can definitely deconsolidate. We saw it back during the Roaring Twenties. We had liberalized markets, and, particularly in agriculture, a tremendous concentration of wealth. Then we had a big stock market crash and the worldwide recession.

"When Roosevelt instituted the New Deal he started with agriculture and we saw a rolling back of that consolidation. We've done it before and we can do it again. It is part of the cycles of capitalism. **The question is how do we usher in a period of reform? To usher in a period of reform you need a countermovement, a social movement powerful enough to force us to institute reforms."**

Maybe the shift is already happening. Philip Ackerman-Leist directs the Farm & Food Project at Green Mountain College in Poultney, Vermont.

"I think we're seeing this 'decentralization' happen right now," he observes. "We're watching citizens all across the United States take on their food systems and challenge the dominant paradigm, this industrialized food system we've inherited. People are suddenly becoming more cognizant and in some cases really angry. This awakening is happening all across the country."

This awakening is certainly happening in Iowa, the metaphoric center of our industrialized food system. If you want to see what single-mindedness looks like on a grand, operatic scale, go to Iowa and witness the most astounding agricultural spectacle on earth. Corn and soybeans cover nearly every square foot of arable land in the state. It's both awesome and unsustainable.

Fred Kirschenmann is a farmer, philosopher, and currently Distinguished Fellow for the Leopold Center for Sustainable Agriculture in Ames, Iowa. He's served on the U.S. Department of Agriculture's National Organic Standards Board and the National Commission on Industrial Farm Animal Production, so he understands the challenges created by our present food system.

"The industrial food system is driven by the same economic principles of any other industrial economy," he points out. "Maximum, efficient production for short-term economic return. Whether you're making automobiles, computers, or food, the way you achieve maximum, efficient production for short-term economic return is by specializing. Henry Ford once famously said that people could buy any one of his automobiles they wanted, as long as it was a black Ford Model T, because that was all he produced. He understood that specializing enabled him to produce automobiles more efficiently. These ECONOMIES OF SCALE allowed him to produce cars at a price that even his employees could purchase.

"It's the same reason we now have these huge monocultures in food production," he continues. "Ninety-two percent of Iowa's cultivated land is in just two crops—corn and soybeans. That's specialization. Simplification of management. You use whatever technologies you can to simplify your management and increase your economies of scale.

"We say we have the most efficient food system in the world. But when you look at it from the point of view of energy inputs and efficiency, it's just the opposite. It now takes ten kilocalories of energy per every one kilocalorie of food we produce. Energy costs will continue to go up as we deplete our resources of fossil fuels."

To learn about alternatives to our present "economies of scale" industrial agriculture, I visit Benzi Ronen. We aren't sitting at a Berkeley café or standing in a Kansas wheat field or mingling at a Milwaukee food security conference. We're at a diner in Tel Aviv, and Ronen explains a principle called ECONOMIES OF COMMUNITY.

As he explains, "If industrial agriculture uses economies of scale to maximize efficiency—focusing on single crops and reducing input costs to a minimum—then communities can leverage their greatest assets—proximity, familiarity, and private ownership—to compete with the global food system. They can create economies of community. The more you decentralize and empower individuals, the better off everyone is. You cut warehousing and retail distribution costs by creating a direct relationship between farmer and consumer. Then you embolden these farmers to become entrepreneurs, self-sufficient companies that know it is good business to develop sustainable growing methods."

THE LOCAVORE

As Jessica Prentice, a Bay Area chef, food writer, and community **KITCHEN INCUBATOR**, observes, "Thousands of years of human history tell us that one of the few things societies need to survive is a sustainable food system—a food system that can sustain itself over time—because when that food system doesn't sustain itself anymore, the society collapses. **It's simple. We have to be able to feed ourselves in a way that's not only sustainable for us as a species but also sustaining for us as individuals. We have to give ourselves the nutrients that we need or we die. We won't die without cell phones, cars, airplanes. None of those things are necessary for our basic survival. Food is.**"

A LOCAVORE gives precedence to food that's locally grown. In many cases this leads the locavore to know who grows his or her food. It's a simple concept and perhaps no other word more accurately defines the act of living in a local food movement. The *Oxford English Dictionary* even made it its 2007 Word of the Year. So, it's a big, important word. Prentice coined it by first looking at the Latin root for "place"—*locus*,

which is how we get words like "local" or "locomotion"—then coupling it with *vorare*, the Latin verb for "to eat" or "to swallow." It's also the root of "devour" and "carnivore." Putting the two roots together gave her *locavore*.

A few years after she originated the word, we're sitting on the front steps of her Berkeley-adjacent home, drinking coffee, when I ask what the term means to her now. She's quick to point out that being a locavore does have limits.

"To me, the idea of being local is not so much about geography or even food miles," she says. "All those things are important, but to me, **the beauty of a local food system is that it brings you back into a relationship with the source of your food, with the land, the animals, the plants, the farmers, and with each other.**"

This idea of relationships has become a rallying cry for many. The USDA even created a marketing initiative around it: "Know Your Farmer," which uses a variety of programs to connect local food producers with their communities.

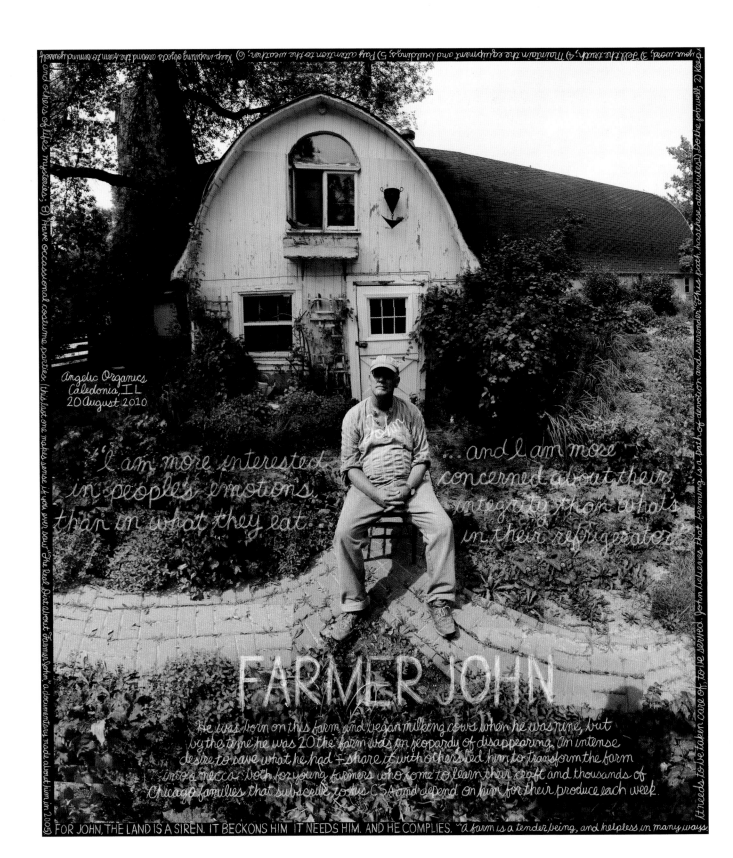

Angelic Organics
Caledonia, IL
20 August 2010

"I am more interested in people's emotions than in what they eat

...and I am more concerned about their integrity than what's in their refrigerator."

FARMER JOHN

He was born on this farm and began milking cows when he was nine, but by the time he was 20 the farm was in jeopardy of disappearing. An intense desire to save what he had + share it with others led him to transform the farm into a mecca: both for young farmers who come to learn their craft and thousands of Chicago families that subscribe to his CSA and depend on him for their produce each week.

FOR JOHN, THE LAND IS A SIREN. IT BECKONS HIM. IT NEEDS HIM. AND HE COMPLIES. "A farm is a tender being, and helpless in many ways

KNOW YOUR FARMER

Knowing your farmer sounds like fun. You might even get to pet a goat or feed a horse sugar cubes from your outstretched palm. Maybe pick your own strawberries. **The fact is, if you don't live in the country, knowing your farmer probably won't happen. Better to start with baby steps, like knowing there are still such things as farmers. If you haven't heard, they're a quietly disappearing breed.**

My grandfather farmed for a time—I learned to drive grinding gears on a Chevrolet Series 3100 half-ton truck in the vineyard behind his Santa Rosa, California, farmhouse—but the lifestyle never took with his two sons. I'm no farmer, either, though I do have the idea to cross this great land and meet farming folks in the places where they live and work, to watch and hopefully learn what a more accountable, more equitable, and possibly even more sustainable food system looks like.

I'm not interested in numbers, charts, and graphs. These quantifiers may provide the illusion of business-like order, but data that can't be seen, touched, smelled, or tasted is ultimately meaningless. Nor am I interested in learning from those who take their directions from the back of a fertilizer sack. Industrialized agriculture is a dead end. It's both extractive in nature—it removes nutrients from the soil without putting anything back—and dependent on pesticides and fertilizers made from natural gas and earthed minerals of limited supply. Nor do I advocate a return to the horse and plow, though it retains a certain charm. The farmers I talk to learn from repeated observation. From errors patiently remedied. From experiences hard won, then freely shared with other farmers.

I travel with a camera, muck boots, and questions. What is the mind-set of these unconventional farmers who grow our food without much government support, who are trying to change the way we eat in this country? What are their motivations? Their values? Why are they engaged in such a physically challenging, financially unstable profession? What's in it for them?

Their answers don't come easily. I might spend an entire day with a farmer, then drive off having never taken a picture. Why? Because I need that glimmer to appear. When people get passionate about their work, light comes to their eyes. After that it's a mad dash. We're on our knees, digging through untilled earth to find white clumps of mycorrhizal fungi. Climbing trees to inspect the structure of budding leaves. Prying bee-hives open. Picking and slicing and tasting. Exploring experimental soil labs housed in leaking storage sheds behind barns. Bobbing in aluminum-hulled dinghies, pulling crab cages out of a bayou. It's hard work, painful and unforgiving. As John Peterson of Angelic Organics in Caledonia, Illinois, tells me one afternoon, resting in the shade beside his tractor, caked with sweat and dust, **"Wanting to farm isn't enough. A person should feel that they have to take it on, because farming is not a destination. It's a path. In many ways it's a path of surrender, of devotion."**

EATING IN SEASON

I arrive in Vancouver two hours late. I blame this on my daughter. It was her idea to bring our dog along for this road trip, which turned our border crossing at Blaine, Washington, into an unexpectedly time-consuming misadventure. I'm here to meet Tyler Gray, the owner of Mikuni Wild Harvest, a unique food company that sells only **FORAGED** food. Everything from truffles to snails to mushrooms to exotic forest greens like fiddlehead ferns and wild licorice root. His is strictly a seasonal business—his company only sells what is found in the wild—and his clients include many of the top chefs in the United States.

We finally rendezvous at a tiny Vancouver café. Gray has promised to reveal one stop on the **FORAGING CIRCUIT**, a network that spreads across Canada and throughout the Pacific Northwest, even as far south as Eugene, Oregon. One such spot is a tiny rock-bound beach near Vancouver's Lions Gate Bridge. If I'd arrived on time Gray would simply wade into ankle-deep water and **WILD HARVEST** a dozen kelp bulbs and their long, semitranslucent leaves or blades, but—as previously mentioned—I'm two hours late. The tide has risen. I stand on the shore, watching sheepishly as Gray strips down to his boxer shorts and dives into the water. It's October and the water is cold. Then it starts to rain. Gray bobs on the surface, then dives repeatedly until returning to shore with an armful of kelp.

An hour later we descend into a thick forest on the far side of Vancouver, pushing through dense foliage to discover mushrooms. Lots of mushrooms. Our wicker basket quickly fills with an assortment of wild goods. There's the "Chicken of the Woods"—called that because it tastes like chicken—and one named the "cauliflower" because . . . it looks like a massive cauliflower. The whole event would seem magical except we're never more than twenty yards off the fairway at a prominent Vancouver golf course. The continuous cursing elicited by one botched tee shot after another becomes the comedic sound track to our foraging odyssey.

I joke that Gray's secret hunting ground will be lost once the first golfer dives into the forest to retrieve an errant golf ball, but he's not so sure. "I can't tell you how many times I've been with someone and picked a patch of miner's lettuce or Siberian oxtails and they're like, 'Wow. I've got that growing in my backyard. I had no idea you could eat it.' **There's this belief that wild foods are dangerous. Maybe it's not as much now, but our relationship with the wild and with wild foods is still in its infancy compared to Asia or Europe.**"

As I spend more time with urban foragers, the narrative repeats. While Gray caters to a sophisticated clientele, the **FALLEN FRUIT** collective literally takes it to the streets. They create richly detailed maps identifying Los Angeles locations where food grows on public property. It might be a tree laden with fruit, its heavy branches extending across an alley. When I join the group on a walk one afternoon in the city's Silver Lake neighborhood, I'm introduced to a row of banana trees growing right out of the sidewalk.

FORAGE:
The art of finding and enjoying wild food.

FORAGING CIRCUIT:
Area where indigenous, wild foods are gathered.

WILD HARVEST:
The collection of indigenous foods; an assembly of native plants, animals, vegetables, fruits, and berries procured from the wild.
—Tyler Gray, Mikuni Wild Harvest

Tyler is very cold right now (and it's my fault)

Vancouver Harbor
Vancouver, BC
24 september 2010

← bull kelp

↑ rain drops

THE FORAGER

Tyler collects wild seasonal foods for restaurant chefs across North America. We agree to meet one morning at 9AM so he can show me the bull kelp harvest. I'm two hours late (border problems) so by the time we get to the water the tide has risen dramatically. Instead of casually picking kelp off the rocks, Tyler must dive into the frigid waters (in a steady downpour). I end up buying him lunch.

Bull Kelp grows all along the Pacific Coast... from Southern California to Canada. It grows year round but the healthiest kelps are found in spring and summer.

WILD HARVEST

The foraging of foods which grow in the wild without cultivation or human assistance

CHEF JOËL WATANABE'S TWICE CURED SMOKED SOCKEYE (WRAPPED IN BULL KELP) WITH HEIRLOOM TOMATO AND SEA ASPARAGUS: mix sugar + salt with zest of 1 lemon, lime + grapefruit + five pcs crushed star anise with 2 tbsp each fennel + coriander seeds. Cover 2 lbs fresh sockeye in mixture and refrigerate 12 hrs. Rinse fish. Place on rack, return to fridge 12 hrs. Remove and cold smoke for 25 minute. Blanch sea asparagus for 10 seconds. Place thinly sliced fish between bull kelp sheets wiped with sake. Cure 20 mins. Whisk plum vinegar, olive oil, mustard and maple syrup and drizzle over sea asparagus salad. Serve.

PEOPLE RARELY EAT THE FRUIT GROWING IN THEIR OWN GARDENS. They simply assume its somehow not as good as fruit from the market. That two block journey...

MINER'S LETTUCE
It has a really great citrus accent, which makes it perfect for salads (add champagne vinaigrette or with roasted beets.) The earthiness of the beets accents the greens very well.

Golden Gate Park
San Francisco, CA
16 December 2010

FORAGE SF
Some of the most prolific wild edibles Iso forages in the city: wild radish greens + flowers, fennel, nasturtium, nettles, miner's lettuce, mushrooms (porcini, shaggy mane) and snails. Crossing the Bay in any direction leads to: seaweed, wild mussels, clams, fennel, mushrooms, nettles, acorns, greens and fruit.

Iso

EATING IN SEASON

Wild edibles grow everywhere. You need to be aware of what's around you. When you spend time outside, see how things change throughout the year.

ISO RABINS EXPLAINS THAT A FORAGER SEES FOOD EVERYWHERE. It's at the park where you walk, in the pond you sit next to, in the trees you walk by, even in your backyard (you just don't know it.) What matters most is being where it's fun. It's how to discover your own foraging spots (because we went there and) and besides, half the fun is the search. Iso's favorite is a rarely used city park. Most people have never heard of it, even though it's in the heart of the city.

"The funny thing is watching people's reactions," David Burns, one of Fallen Fruit's three founders, tells me. "Most people in urban areas have never seen fruit in the wild, even when it's growing on their street. **You can pick one of these bananas and hand it to them, but if it doesn't come from a store and isn't wrapped in plastic, they just won't eat it."**

Iso Rabins founded ForageSF to support a nascent community of urban food explorers. His favored haunts include San Francisco's public parks, with these grounds proving so fertile that fellow foragers source its wild ingredients to make artisanal foodstuffs and later sell them at Rabins's underground food markets. **Foraging awakens people to the seasons and what's around them. It's the same for people who grow or buy fresh food.** February means—at least where I live in Petaluma—that chickens start laying eggs again. Fava beans arrive, along with nettles. The first raspberries miraculously appear. Blackberries come right after. A rolling tide follows of bush beans, beets, fennel, chard, Padrón peppers, snap peas, tomatoes, and kale. Always kale. Ditto for zucchini. You eat these things in mad flourishes, until you're tired of them. Then the next year rolls around, you faintly remember what you missed, and the cycle repeats. These foods remind you of the passing seasons. They become something to look forward to.

I sometimes have the same jarring recognition when finding myself outside at night, looking up and suddenly taking notice of the stars above my head. That twinkling canopy has always been there. At one time people used those same stars to tell time, guide their journeys, and explain the heavenly mysteries. Now when we look at the moon, we don't know whether it's waxing or waning. Nor do we know from where it rises or sets. Our lives are full of these opportunities, these invitations to check in with the natural world, to reaffirm and reconnect.

Or we can simply cheat nature, bending it ever so slightly to our will. It's clear that some consumers will only take this "eating in season" business so far; they'll still want their tomatoes in January. At that time of year these foods are definitely local for somebody, but not *you*. **SEASON EXTENSION** means using greenhouses to prolong the growing season. Seeds are planted earlier and the last harvest of the year comes later, allowing consumers to continue eating local foods longer than would otherwise be possible.

On a typically foggy May morning in Washington's Snoqualmie Valley, Andrew Stout of Full Circle Farm in Carnation, Washington, explains this principle. "We can educate people," he says. "We can explain that seasonality means different produce appears at different times of the year, but to compete with your local grocery store we have to be a year-round business. We have to grow things people expect to find, regardless of the season."

He's right, but what if I took his "greenhouses empower small farmers to satisfy consumer needs" theory to its furthest extreme? What would growing everything all the time look like? I get a glimpse some time later, when touring tomato greenhouses in the Netherlands. These red globes are known to their German neighbors as *wasserbomben*, or "water bombs." And the Germans are right. When no one's looking I bite into one. Predictably, it has no taste. Where is today's harvest of "water bombs" bound? An American supermarket near you. We want our tomatoes twelve months out of the year and fully robotic vehicles dragging trailers loaded with tomatoes are happy to comply. I follow them through a maze of glass-paneled greenhouses lined with acre after acre of tomatoes. The plants are ten feet tall, their branches—laden with fruit—spread across thin metal cables. The base of each plant sits inside a small cardboard box; there are practically no roots. Every ounce of energy generated by this plant is dedicated to the production of fruit. It's wickedly efficient.

Is this merely season extension or have they forgotten seasons entirely? These producers are capable of growing any vegetable at any time of the year. They might not taste like much, but they sure look like the real thing. Besides, if you eat them long enough, you'll eventually forget what a real tomato tastes like.

IS ATHENS, GEORGIA, THE NEW LOCAL?

No single definition explains what it means for food to be local, because the concept is surprisingly vague. Somebody might say "local" is defined by distance—using food miles as a measure—or by boundaries, using a specific town or county. Or, as the director of an Iowa food co-op once told me, pointing to a chalkboard map she'd drawn, "Local is anything grown in the entire state." It might be framed according to geographic considerations, by valleys and rivers and mountains. Or it might be defined by cultural ties, like a far-flung community of immigrants united by a common language and diet.

Clearly, when talking about things local, context is everything.

When you take a major river or waterway, determine all the water sources flowing into it, then trace these back to their points of origin, that total landmass is a **WATERSHED. Watersheds come in all shapes and sizes and are oblivious to arbitrary boundaries. The only law they respect is that of gravity. Water flows downhill.**

A **FOODSHED** respects another set of laws, one chiefly defined by economics, by supply and demand. The term also hints at a community's ability to feed itself. There was a time when settlers lived off their land; they ate what they grew. Modern culture is more complex. Take any package off your pantry shelf. Its ingredients probably come from a dozen locations around not only the country, but the globe.

Now imagine a **LOCAL FOODSHED**. It's a visionary exercise, one that's admittedly quixotic, the idea that communities can still feed themselves. We live in a world of infinite choice, and our diets reflect that. People want what they want, when they want it. Still, what would a local foodshed look like? First, it wouldn't merely occupy physical space. Nor would it be measured by distance. It would chart a web of personal relationships—**KNOW YOUR FARMER**—and it would ebb and flow depending on the time of year—**EAT WITH THE SEASONS**. Last, it would be linked by infrastructure—how things are transported from field to plate. If you put all this on a map, an alternative food system appears. Call it the new

Famworth/Bershire cross bred (4-7 months)

CREA
FULL
FOO

Jason moved he
"The Bay Area is
this country's foo
he recognized a
The people of Athen
and what they ate
thriving food scene,

THE NE
A system of localized re
(fundamental to

JASON MANN SAYS "AS A SOCIAL AND ECOLOGICAL ENTREPRENEUR, I BELIEVE I

top margin (handwritten, inverted): of to sustain their formulation and those in low-income communities having able to cultivate healthy + conscientious eating habits + consumers living able to have access to these products. It's about farmers being about EVERYONE, not about

upper right (handwritten):
Moonshine Meats
Athens, GA
29 June 2011

Jason

Jake

FULL MOON FARMS
(a vegetable farm + CSA)

2. FARM 255
(locally and seasonally-sourced restaurant)

3. MOONSHINE MEATS
(collective of partner-farmers producing
pasture-raised livestock: they create their
own fertility + raise all proteins for restaurants)

4. MEAT CSA
(cow + pig for
local subscribers)

5. Assistance to develop
local slaughterhouse

6. FARM BURGER
(casual restaurant with lower price point yet
still using ingredients from their own meat
cooperative and vegetable farm)

left margin:
OCAL,
GRATED
NESSES
n Berkeley, CA
tle crescent of
ement") because
nd a challenge.
about the land
didn't have a
elped create one:

N FOOD ECONOMY

nships which defines how our food travels through the supply chain
equation is the welfare of humans, animals and agriculture)

right margin (vertical): It's not just about pasture-raised pork or locally sourced vegetables, but about EVERYONE having access to these products. It's about farmers being

bottom: VER OF THOUGHTFUL BUSINESS MODELS. THEY PROVIDE A BENEFIT THE PUBLIC SECTOR CAN'T COMPETE WITH IN THE REALM OF CREATING SOCIAL TRANSFORMATION.

local, where a community's inhabitants stop waiting for the government to fix their food system and just do it themselves, one piece at a time.

Creating a food map shows you where to start. As Philip Ackerman-Leist points out in his book *Rebuilding the Foodshed*, "It's not just about identifying the particular production, processing, or distribution [in a region]. It's a much more complex cycle. You start to see flows that are there . . . and certain ones that aren't. Very often those gaps in a foodshed have come about by virtue of the industrialized food system we've inherited and the fact that we specialize our food systems to such a degree.

"We've got areas that may be incredibly rich in terms of what they're producing—in many cases corn or soy, those kinds of commodity crops—but there's this incredible dearth of diversity. That's really what this is about: re-instilling diversity and allowing it to flourish. It's a long-term cultivation act. It's not something we just do then walk away."

What does a "long-term cultivation act" look like, one that "re-instills diversity" in a community? Jason Mann knows. He even has a name for it: the **NEW FOOD ECONOMY**. He lives in Athens, Georgia—a liberal mecca set in a southern football-crazed college town that still flies the Confederate flag—but he isn't from there. Mann recounts his "Road to Damascus" moment as we pursue a dozen Tamworth/Berkshire cross pigs through acres of dense forest just outside town. **He recalls being deeply committed to the San Francisco League of Urban Gardeners until having a sudden realization: San Francisco probably didn't need another local food advocate.** Months later, on a cross-country trip, he made a fortunate pit stop. "Athens at the time . . . there were no farmers' markets, literally zero CSAs, and no farm-to-table restaurants where people came together to celebrate and eat and support people doing that work," he remembers. "A handful of farmers worked in a vacuum, doing important work, but they weren't really communicating with one another. On the nonprofit side, Georgia Organics, a regional organization promoting organic practices, was just getting its legs. **All the things we've come to appreciate in California—the social fabric, the network, and the communication—none of that was happening here.**"

What Ackerman-Leist defines as "gaps" in a food system, Mann and his partner, Olivia Sargeant, saw as opportunities. They started with a farm that was **BIODYNAMIC**, coaxing **SOIL FERTILITY** out of tough Georgia clay that, like most of the Piedmont, has suffered mightily after a century of hard soil mining. They grew biodynamic produce, then started a CSA. Their farm-to-table restaurant, Farm 255, came after.

"It was a challenge," Sargeant says. "Before we started Moonshine Meats it was really impossible to source local meat. We were committed to not sacrificing our whole animal mentality or our commitment to sustainable animal husbandry, so there was a lot of driving around Georgia and a lot of skirting loopholes, a lot of meat packed in coolers and a lot of trouble."

Inevitably Mann and Sargeant decided to raise their own cattle and hogs. Chickens came after. "It really made sense, especially sitting on one hundred acres of pastureland that was already designated for grassland restoration, to start an animal program and build that fertility," Sargeant says. "From there it obviously grew to including other producers to meet the volume we created."

Setting up Moonshine Meats, a local meat cooperative and CSA, created a network connecting local producers with consumers. It also led the duo to confront another one of Ackerman-Leist's "gaps" in the food system: the lack of local processers. They found a solution for that as well by striking a deal with West Georgia Processing in neighboring Carrollton, which slaughters their animals then breaks them down, saving Mann and Sargeant hours of driving and processing time each week. Sargeant has a name for rebuilding this infrastructure on a community level: **RELOCALIZING**.

Establishing local meat production and securing processing opened up additional opportunities, what Ackerman-Leist defines as flows. Mann and Sargeant soon started their own version of a fast-food chain: Farm Burger. It features their own antibiotic- and **HORMONE-FREE** grass-fed beef, with the patties loaded with locally sourced arugula, fried eggs, cured lardo, and vine-ripened tomatoes (gluten-free buns are available). In addition, they have their own coffee roaster and

"Let's bring back the guilds, the granges, the purveyors, the merchants and the artisans so we have both craft and community." - Olivia

Olivia and her partner[1] wanted a local slaughterhouse[3] to process meat[3] for their two restaurants[4] and a meat CSA[5] (instead of driving 300 miles a week)

Tim[6] opened this facility

(his childhood dream) to become more involved in his community and help his friends[6] — be more sustainable.

1. Jason Mann[6] first discovered Tim[6]
3. Moonshine Meats is their pasture-based cooperative
4. Farm 255 (Athens, GA) & Farm Burger (Decatur, GA)[7]
5. 18 subscribing members

RELOCALIZE

to revitalize a community by favoring local food[3], goods[7] and people[6,8,9]

West Georgia Processing[2]
Carrollton, GA
25 June 2011

OLIVIA SAYS "OUR FOOD SYSTEM IS SICK AND OUR MEAT IS DIRTY. PART OF THE REMEDY IS TO LOOK INWARD AND REINVENT LOCAL. Even taking the smallest of

a food truck business, the inevitable appendages of any vertical food business operating in the New Food Economy. Demand triggers Inspiration. Capability increases Capacity. Profit encourages Expansion. Then the cycle repeats, with others being inspired to follow their example. This is how local food systems are built.

PIG HEADS
(they try to use all parts of the animal, especially as condiments made from bone marrow, ox tail marmalade, house-cured lardo, house-cured guanciale, pork belly, beef cheeks, pork cheeks + pickled beef tongue)

WHY THIS ISN'T YOUR TYPICAL FAST FOOD HAMBURGER:
1. Their own producer cooperative, Moonshine Meats, rai
2. These animals are never given antibiotics, growth horm
3. The whole animal is used, with the exception of p
4. Their beef is handled with passionate hands - from far
5. Farm Burger purchases its other ingredients from loca

Farm Burger
Decatur, GA
25 June 2011

HOW TERRY KOVAL MAKES A FARM BURGER: 1) Gently form a ball of grassfed beef 2) Season with kosher salt and cracked black

Vidalian smoked cheddar on the burger; 10) Place on bun and top with the caramelized onions and Farm Burger sauce

celery, carrot, onions, herbs, chili garlic sauce and spices); 8) Butter bun and toast golden brown; 9) Melt a slice of aged

GEORGE
(co-owner)

TERRY
(chef)

"It is all about the ingredients, the people who provide them and the places they come from."
—Terry

(SLOW) FAST FOOD

...nishes) their animals on a diet of local grass (no corn)
...teroids
...s (tenderloin, strip loin + rib-eye go to their sister restaurant, FARM 255 in Athens, GA)
...rocessor to butcher to chef to cook
...(keeping money in the community where they live + work while reducing their carbon footprint)

...n a hot griddle add oil + ball of beef; 4) Use a spatula to press burger into a patty (this will give the burger a crispy outside);

...fresh thyme, salt and cracked black pepper until they are a dark, sweet caramel (Loves.) 7) Prepare sauce (Farm Burgers is made of...

5. Turn the burger over and finish to a juicy medium. Done. 6) Slice sweet Vidalia onions (local) and caramelize them on the griddle with

BUILDING INFRASTRUCTURE, or HOW THE ROTO-FINGERS® PEA-BEAN SHELLER WILL SAVE THE WORLD

The first casualty in the centralization of any food system is local processing. Local dairies left without bottling plants. Poultry producers forced to travel several hours to the nearest slaughterhouse. Wheat growers first losing seed-cleaning operations, then local grain mills. Rebuilding local production capacity means nothing when there's no way to process things after they're grown.

In Sheeplo, Mississippi, I visit a packing shed. It's a sheet metal warehouse with a concrete floor. I'm out of the sun but the noonday heat and humidity are oppressive. I feel like I weigh five hundred pounds.

I've come to see something called the Roto-Fingers® Pea-Bean Sheller. This ingenious device demonstrates another way to rebuild local food systems. Ben Burkett founded the Indian Springs Farmers Association, a cooperative of local farmers based in Sheeplo. He purchased the sheller and installed a walk-in refrigerator, all to provide cooperative members with the costly processing equipment needed to compete with larger producers. The Roto-Fingers® Pea-Bean Sheller is especially useful. It shells a bushel of Pinkeye Purple Hulls in under ten minutes. A bushel is twenty-eight pounds of peas. Shelling them by hand would take hours—precious time Mississippi consumers would rather devote to other endeavors. They're willing to pay a premium—which for a farmer means profit—for shelled peas like the Purple Hull. Having this sheller creates new markets, incentivizes growers to plant more crops from their childhoods, and keeps local traditions alive.

Stories like these repeat across the country. A few months later I travel north of Seattle to the San Juan Islands. Lopez Island is remote and only accessible by ferry. Its rolling hills are predominantly grassland, ideal for raising cattle, except when it comes to processing. How do you get these beasts off the island? Confronting this problem set local farmers and the Lopez Island Land Trust on a quest that took six years. They built a USDA-regulated **MOBILE SLAUGHTERHOUSE** inside a steel-walled trailer pulled by a semi.

Nick Jones, a cattle farmer on the island, will process ten animals this morning—three-year-olds he's grazed on the low hills that stretch in every direction. The mobile slaughter unit (MSU) is already there when I arrive. It's parked in a clearing beside the barn.

It's a gleaming metal box. Every surface shines. A butcher and two apprentices are already at work. There are knives to sharpen, tools to prep. While they finish setting up I walk the farm with Jones. He's tall, razor thin, and possesses an austere demeanor. We stop at his impressive vegetable garden—just for the family—then peek inside a tiny farmstand that sells his pork, beef, and shellfish. Shellfish? It turns out Jones is also a fisherman who farms clams, oysters, and mussels.

Indian Springs
Farmers Association
'Sheeplo, MS'
Downtown Hattiesburg Farmers Market
APRIL - OCTOBER · THURSDAYS · 3 PM UNTIL 6 PM

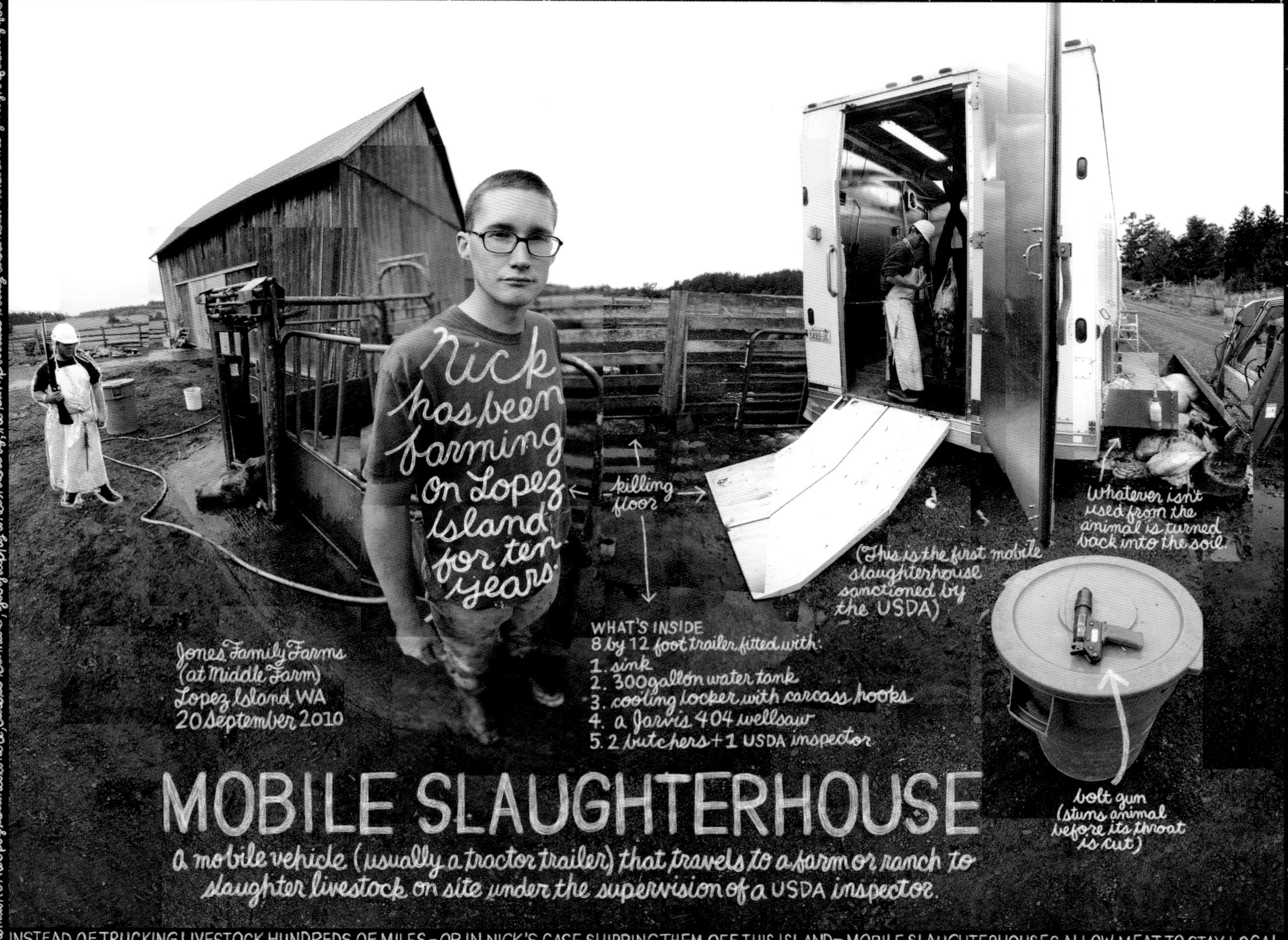
Nick has been farming on Lopez Island for ten years.
killing floor
(This is the first mobile slaughterhouse sanctioned by the USDA)
Whatever isn't used from the animal is turned back into the soil.
Jones Family Farms
(at Middle Farm)
Lopez Island, WA
20 September 2010
WHAT'S INSIDE
8 by 12 foot trailer fitted with:
1. sink
2. 300 gallon water tank
3. cooling locker with carcass hooks
4. a Jarvis 404 wellsaw
5. 2 butchers + 1 USDA inspector
bolt gun
(stuns animal before its throat is cut)

MOBILE SLAUGHTERHOUSE

a mobile vehicle (usually a tractor trailer) that travels to a farm or ranch to slaughter livestock on site under the supervision of a USDA inspector.

INSTEAD OF TRUCKING LIVESTOCK HUNDREDS OF MILES – OR IN NICK'S CASE SHIPPING THEM OFF THIS ISLAND – MOBILE SLAUGHTERHOUSES ALLOW MEAT TO STAY LOCAL.

The MSU allowed Nick to incubate his farm in the early years. While it isn't the most economical solution (it costs approx. $650 per cow, $250 per hog and about $120 per goat/sheep), the MSU connects Nick with his fellow farmers and allows him to play a vital role in his community. The MSU also allows Nick to remain on Lopez Island where, due to climate, geography and history, he can produce truly distinctive extremely high quality foods.

The cattle wait in a paddock behind the barn, out of view. They're loud, more curious than agitated by the unexpected break in their daily feeding routine. Jones leads them one at a time to an open area beside the trailer. He talks to them, calls them by name, whispers and rubs their shoulders. Minutes later they're hanging inside the truck as Jones disappears behind the barn for the next cow.

It's brutal and efficient, yet oddly humane; conundrums like these are typical of life on any farm. The animals are stunned senseless, bled out, then dragged into the truck. Once hanging they're opened up and their organs are removed. These are pushed out a side door, creating a steaming, viscous pile that Jones disposes of with his tractor. The offal will be turned into the soil. Next spring vegetables will grow on top of it.

Back inside the truck, the hanging carcass is halved with the industrial chop saw, then quartered, with each piece firmly hooked to metal rods hanging from rails bolted to the roof. This allows the sides of beef to be easily slid toward a refrigerated unit at the front of the truck.

The USDA inspector mainly remains in his car. He's got a newspaper, a thermos, and classical music on the radio. He's been doing this a few years now. Everybody knows everybody. Rules are followed. Etiquette is obeyed. The whole affair winds down in a few hours. Paperwork is signed and stamped, then everyone goes on with his day. By this time tomorrow the meat will be in a dozen restaurants and small markets both on Lopez Island and in nearby Seattle.

How does he find the time? "They're both satisfying, honest pursuits which put you in beautiful surroundings and force you to deal with the physical world on its own terms—not yours," he explains, while idly pulling onions from the ground. He taps each one gently against his thigh, watching as hard clumps of dirt fall from the roots to the ground. The butchers are waving their arms. It's time to start.

I notice a white sedan parked off to the side, under some trees. A man dressed in a white lab coat gets out. He's holding a clipboard. He's the USDA inspector, the single thing that makes this morning so utterly amazing. This is an inspector not at an industrial slaughterhouse on the Great Plains, but in a field, on Lopez Island, standing inside a steel-walled tractor trailer.

David Evans raises two types of chickens: layers and broilers. The first produces eggs. Their productive lives may extend to four or five years. A broiler is what you eat, with the Cornish Cross being the preferred breed. In industrial agriculture a broiler reaches slaughter weight in six weeks. Artisanal poultry producers often **PASTURE-RAISE** their birds and provide organic feed. That and the different breeds used add another four weeks before slaughter.

Evans's birds are processed in the front yard of his farm in Inverness, California. A bird is placed in a steel cone, with its head protruding through a hole at the bottom. Its throat is cut and the bird bleeds out. Then the bird is dropped into a bath of scalding water. This loosens the bird's feathers for what comes next: a washing machine lined with rubber fingers. These spin, deftly removing feathers. The bird is then gutted, dressed, and placed in an ice-water bath.

Evans's crew will process a few hundred birds in this fashion before noon. These birds will appear twenty miles away at a San Francisco farmers' market the next day and sell out within hours. By eliminating trucking and using on-site slaughtering, which puts less stress on an animal, producers are able to reduce costs and keep things local.

I return to Evans's farm two weeks later, only to find that inspectors from California's Department of Food and Agriculture have shut Evans down. Their reasons

are many, all stemming from the fact that he's a victim of his own success and has too many birds. Evans now has to drive his poultry to Sacramento, more than two hours away, for processing.

He encounters similar problems processing his beef. Not owning a facility makes it hard to build his business. Then Evans achieves the unimaginable: He opens his own USDA-regulated processing facility in nearby San Francisco. This serves not only his needs but those of other cattlemen in the area, adding yet another piece to the local food system puzzle.

There is little difference between the Roto-Fingers® Pea-Bean Sheller, the Lopez Island mobile slaughter house unit, and Evans's meat processing plant. Each uses ingenuity to solve a very real problem, that of finding new ways to produce food on a local level.

FARMERS' MARKETS

A few years ago, my wife, Laura, had an idea. She borrowed a few steel milk jugs from a local dairy, then carted them over to a goat farmer who lives five miles down the road. She mixed his fresh goat's milk with a few fanciful ingredients like figs, balsamic vinegar, molasses, espresso, and a complicated cabernet wine reduction to make ice cream in her Petaluma kitchen. It was an experiment. **Laura had worked nearly fifteen years as a producer in Hollywood. Now she was ready for something new. But what did she know about starting a goat's milk ice cream company?**

A local scoop shop let her use their equipment after hours. She hand-packed the ice cream in white pint-sized cups. Labels were affixed by hand, one at a time.

Where did she sell those first pints? At our local **FARMERS' MARKET**.

Farmers' markets are moveable feasts that mean local and fresh. They epitomize the principle of **SHORT SUPPLY CHAINS**—the distance between producer and consumer. The vendors are a mix of local farmers. Farmers' markets are also a business incubator for artisans selling everything from handmade mozzarella to ethnic delicacies from countries across the globe. They give nascent producers instant access to new customers, which is vital. The success of these producers is critical to rebuilding local food systems.

Laura's Sunday mornings began with loading the Laloo's ice-cream truck at 5 a.m., then driving to a patch of dirt beside the local courthouse. The vacant lot was filled with vendors by 7 a.m. You wouldn't expect to sell much ice cream at that hour. In fact, she often brought the Sunday paper, but it never left the front seat of the truck. Farmers' markets are about conversations. Even before she'd set up her tent people would be at the table asking questions. "Does it taste like goat?" "Is it better for you than cow's milk?" "Can I have it if I'm lactose intolerant?" Laura started conducting impromptu science lessons each week, using diagrams that showed the difference between lactose molecules and caseins in milk from goats and cows. After all, a farmers' market vendor is part salesperson, part educator.

Customers also brought stories. The most unlikely people were raised on farms. They told tales of milking goats, of chasing them through fields. Raising them with a bottle. **Farmers' markets remind people of an earlier time when they were closer to the land and closer to the people who produced their food.**

After her ice cream began appearing in local stores, many customers still preferred buying it on Sunday mornings. It was the same product, but the one from the market came from a person they now knew.

Researchers and academics like to study farmers' markets and journalists like to write about them. Some say they build vital local infrastructure. Others say they're a fad that will eventually wane as consumers tire of adding another burdensome stop to their weekly shopping list. Then comes the idea that farmers' markets are elitist, that their mostly organic produce is sold at prices higher than a local supermarket, while their cash-and-carry business model excludes people on federal assistance, because farmers don't take food stamps.

Actually, many do, except they're not called "food stamps" anymore but EBT. Individuals enrolled in **SNAP** (the Supplemental Nutrition Assistance Program) are given EBTs (electronic benefit transfers). I travel to New Orleans to see how farmers' markets turn these EBTs into cash. A group called Market Umbrella operates a number of area farmers' markets. They've pioneered a wireless transaction system that allows SNAP enrollees to not only convert their EBTs into wooden coins redeemable with market vendors but even use grants from nongovernmental organizations (NGOs) and state agencies to double the value of these food purchases. It's a mechanism that provides greater food access to a new category of consumer while putting more cash in local farmers' pockets. Richard McCarthy, Market Umbrella's founder, explains it this way, "Many of these people are vulnerable. They have health issues. Integrating SNAP into farmers' markets is critical because these farmers are teachers. They can help educate about nutrition."

Farmers' markets are drafting new unwritten social contracts with their communities. By sharing their knowledge and experiences, farmers build trust, all by selling fresh produce. McCarthy points out that shopping at a grocery store limits you to a single interaction, and it's with a cashier. The conversation: "Paper or plastic?"

"A farmers' market is the complete opposite," McCarthy says. "It requires you to speak with everyone. Buying a head of lettuce practically demands a conversation. The same with apples or carrots. You might learn a new way to cook something or what's about to be in season or how things are grown. Your shopping bag becomes a collection of stories."

51

COMMUNITY-SUPPORTED AGRICULTURE

I'm in Boulder, Colorado, to photograph the OGs, or "Original Growers," who jump-started Colorado's organic movement. I spend two weeks profiling a likable group of farmers on both sides of the Rockies. Peach pickers in Hotchkiss. Mushroom growers in Fort Collins. Beekeepers in Weld County. Then I meet Anne Cure. She's a tall, engaging woman with luminous eyes and deep auburn hair woven into a ponytail that hangs well below her shoulders. Her hands are deeply callused, creviced really, and stained with dirt from working an organic farm right outside Boulder's city limits. Relatively new to the area—having apprenticed for a time with Andrew Stout at Seattle's Full Circle Farm (see p. 38)—she has come to develop a fully integrated farm, mixing livestock with organic produce grown using mostly **BIODYNAMIC** practices. It's a far-flung operation, one made even more impressive when I visit her on a Wednesday afternoon. Cure has a **CSA** (community-supported agriculture), and today her front yard is jammed with cars. These are people who've come to cash in on their investment, an investment made in Anne Cure.

CSAs change from farm to farm, but the concept remains the same. At the start of each growing season, members purchase a subscription. Each week they get a box of fresh produce containing whatever happens to be growing on the farm. That influx of cash at the start of the season allows the grower to purchase seed and farming implements, even hire workers. Essentially, a CSA subscription is a contract between a consumer and a farmer. In a good year a CSA box will contain more produce for a longer time span. Conversely, if there's a drought, a blight, or a crop gets frosted out, the consumer shares the farmer's loss.

Today, Cure's visitors tote canvas shopping bags and cardboard boxes. It's a leisurely affair. The people all know one another. The women exchange hellos while their children run wild, climbing on tractors, kicking up dust, and chasing chickens. Meanwhile, Cure is omnipresent. She explains recipes to one woman, sorts lettuce with another. This is what a community looks like.

The CSA is direct marketing at its most eloquent. Consumers not only buy right from the farmer, but actually invest their money, and part of themselves, in the transaction. It's their farm now. Their farmer. Their reconnection.

Cure's CSA model also offers a valuable mathematics lesson. She produces food for 175 families. In 2010, the U.S. Census Bureau set the size of the average American family at 2.59. So 175 families x 2.59 people = 453.25 people served by a CSA of Cure's size. The same census set the U.S. population at 308,745,538. If one farm with 175 member families serves 453.25 people, and if 681,181 farmers produced as much food as Cure, the entire nation would have access to fresh, local produce for much of the year.

I find CSAs wherever I travel. People complain about getting too much kale in August—I know the feeling—but through CSAs consumers express their desire for a better food system. Why wait for the current retail distribution model to evolve and give you what you want, when you can forge your own relationships directly with a producer? And it's not just with fruits and vegetables. Farmers have created **MEAT CSAs** that allow consumers to receive cuts of grass-fed protein wrapped in butcher paper each week. Want beans or fresh-milled flour from a local farmer? There are CSAs for that. Seafood from local fishermen? That, too.

Anne and her husband are part of the "new face of farming", meaning they are new to the area and have no land history here (though they are the fourth generation of farmers on this land)

This is a 1949 Allis Chalmers model G. They stopped making them in 1953 but for Anne there is nothing comparable for small scale vegetable production. Its rear engine design allows the driver to see everything they are seeding or weeding as it is happening. (all is that history)

FARMER ANNE

As the face of agriculture changes and smaller family run farms return, more and more women are coming into agriculture as their life's work.

ANNE SAYS: "EACH DAY THAT I AM IN THE FIELD, I KNOW THAT I AM JUST ONE WOMAN FARMER OF THOUSANDS ALL OVER THE WORLD SOWING SEED, IRRIGATING, CULTIVATING CROPS,

(top margin, inverted) ...tions and create a place people feed a part of... We coming to feeding so many families at our land can...

(left margin, vertical) healthy provide for. This includes over a hundred varieties of vegetables and flowers grown during the year."

"We want to connect with these families and provide a place where they can experience the farm."

"Our goal is not food for the world. It's to g really good for 175 fami

CURE ORGANIC FARM CSA

SMALL   MEDIUM   LARGE
1. Red Beet
5 Sweet Corn
Summer Squash
Garlic
Cucumbers 5pcs
1 lb. Green Bean
4 Tomatoes

1.
CSA subscribers choose between three box sizes

They directly engage farmers who can share the stories behind everything grown on the farm

Anne

JEFF

TRACEY

NANCY

COMMUNITY SUPPORTE

Buying a CSA membership means entering into partnership with a local farmer. The member pay for seed, water, equipment and labor in the early season when farm expenses are high and fo produce each week. CSAs build community by reconnecting their members to the seasons and fostering re

I ASK ANNE CURE WHAT LED HER TO CREATE THIS CSA FOR 175 LOCAL FAMILIES AND SHE SAYS

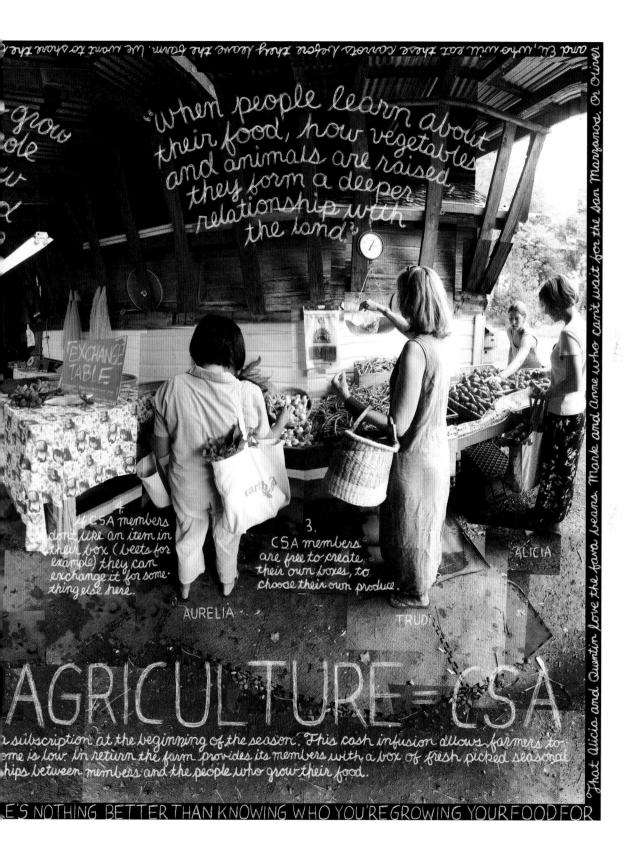

"When people learn about their food, how vegetables and animals are raised, they form a deeper relationship with the land"

EXCHANGE TABLE

4. CSA members dont like an item in their box (beets for example) they can exchange it for something else here.

3. CSA members are free to create their own boxes, to choose their own produce.

AURELIA

TRUDI

ALICIA

AGRICULTURE = CSA

a subscription at the beginning of the season. This cash infusion allows farmers to ...ome is low. In return the farm provides its members with a box of fresh picked seasonal ...hips between members and the people who grow their food.

...E'S NOTHING BETTER THAN KNOWING WHO YOU'RE GROWING YOUR FOOD FOR

A nna Larsen operates a **CSF** (community-supported fishery) in Northern California. She maintains close contact with area fishermen, meets them directly at the boat as they offload their catch, then cleans and packages seafood in ice chests for delivery throughout the area.

Benzi Ronen, founder of Farmigo, a company offering web solutions for CSAs, says we're entering a new age. "We're seeing an economy that's now a *sharing* economy. Everybody is becoming not just a consumer but also an active participant. It's **COLLABORATIVE CONSUMPTION**. Instead of going to a hotel, you can rent somebody's

house. . . . We're going to see the same thing happen with our food system. You can be both a consumer and an active participant in helping sell local food to your neighbors. That will enable the movement to scale."

Some cattlemen now offer a slightly different model of collaborative consumption called **COW SHARES** or

COW POOLING. Consumers gather in groups, use their combined buying power to purchase an entire animal, then after the animal is slaughtered divide the cuts among one another.

Altgeld Gardens Liquor
Altgeld Gardens, IL
18 August 2010

MSP

DISHWASHER SOAP   CHARCOAL FLUID
CRAYONS ENVELOPES   MORE TISSUE PAPER
TISSUES LIGHT BULBS   DOG FOOD CAT FOOD

LAUNDRY DETERGENT

DIAPERS

SODA   SODA
        CHIPS
BEAUTY PRODUCTS   CANDY
MORE BEAUTY PRODUCTS   SOAP
(HERE TO)   SHOE POLISH TOILET PAPER

CANDY
CANDY
COOKIES

Altgeld is one of the oldest public housing
projects in the United States. Aside from
a liquor store situated on site, there are no
nearby food sources for the community's
2,500+ residents.

corner store   115th
119th St.
metro   Michigan avenue
station
123rd St.
gas station
127th St.

FOOD DESERT

An area where residents lack access to affordable fruits and vegetables,
whole grains, low-fat milk, legumes and other food that constitute a healthy
diet. Grocery stores are either inaccessible to these shoppers (due to high
prices and/or inadequate public transit.)

As a result, residents buy food and drinks from gas stations, fast food restaurants
and corner stores which primarily sell processed food. This often leaves these individuals
at risk for obesity, diabetes, and chronic illness.

According to Growing Power's Michelle Lee, "It's a food drought with less food or no food and lots of l

(food desert map of Altgeld Gardens by Teri Payton + Cynthia Beal)

STORE TYPES:
1. corner stores
2. independent + local chains
3. discount chains
4. full service supermarkets

ALTGELD GARDENS
you are here

FOOD DESERTS and CORNER STORES

Wayne Roberts, former director of Toronto's Food Policy Council, tells me that while hunger and undernutrition affect some two billion people around the world, food availability is not the barrier to ending hunger. Instead, the real challenge is **FOOD ACCESS**.

"Food access is determined by a variety of factors," he says. "The income of people experiencing hunger, the racial or cultural background of certain populations, and the distance between people and food markets all contribute to food insecurity. As a result, people have developed approaches to promote neighborhood-based food retail outlets or community gardens in disadvantaged communities, and public education campaigns to highlight such inequities as the prevalence of low-quality corner and convenience stores in underserved communities."

When people talk about the food challenges facing low-income urban communities, areas without supermarkets or neighborhood grocery stores, where nutritious food is scarce or nonexistent, they often use the term **FOOD DESERT**. It's a colorful, descriptive term. It's also inaccurate. **Food deserts are just as likely to exist in rural areas as in cities; anyone driving cross-country knows that. Our centralized food system is partially responsible for these geographic gaps.** Their omissions are less the result of careless oversight and more the outcome of judiciously considered design; either these areas aren't profitable enough or they are too dangerous or inconsequential to worry about, which describes my visit to Chicago's South Side.

Erika Allen, executive director of Growing Power's Chicago office, takes me to Altgeld Gardens, one of the oldest housing projects in America. A young community organizer named Barack Obama got his start here. Years after he left, it's still a food desert with nearly four thousand inhabitants. Practically the only food available for five miles in any direction comes from a single windowless cinder-block building with low ceilings. Its products are displayed behind Plexiglas walls. You'll find dishwasher soap, crayons, tissues, light bulbs, laundry detergent, dog food, cat food, a myriad number

of beauty products, shoe polish, candy, cookies, diapers, potato chips, five shelves of alcohol, and lottery tickets but not a single apple or a carrot.

Allen's going to change that. She shows me her armory, a room full of shiny wheelbarrows and shovels, then we cross the road to watch dozens of teenagers tear up the hard ground. While their first harvest a few months later may seem symbolic, it's not. For these people, the sheer act of growing their own food is transformative.

While growing food on urban farms is noble, it won't feed a city. Enlightened urban planning is required, and that begins with designing communities with more equitable food access. Most people residing in food deserts rely on public transportation. If they can't walk to a market or get there by bus, they'll shop at whatever store is closest. That could be a convenience store or a gas station. Even a liquor store. **A diet based on such impoverished food choices leads to predictably scary health outcomes. Hypertension, obesity, or diabetes: Take your pick. Food deserts have them in ready supply.**

Revitalized **CORNER STORE** projects—which transform liquor stores with limited food options into those offering healthy and even local food—are now underway in urban communities across America. In Portland, Oregon, I spend time at the Village Market with Amber Baker. Village Market serves residents of North Portland's New Columbia housing project. You won't find lottery tickets, alcohol, or cigarettes—the traditional profit centers of any urban market—at this corner store. Nor are the walls festooned with garish posters of bikini-clad women hawking cold beer. These have been replaced by the "Food for Thought" wall. This week it celebrates blueberries with a recipe for whole-wheat blueberry muffins and a handy blueberry nutrition guide ("very high in antioxidants" and "low in calories"). To the right, a brightly worded factoid about carrots proclaims, "After you cook carrots, the amount of absorbable vitamin A doubles!" **At the new corner store, it's not enough to paint the walls in bright colors. You also educate consumers. If they don't know how to cook or eat nutritiously, it won't matter what's on the shelves.**

(the new) CORNER STORE

Providing healthier and more economical food choices for consumers living in urban communities

Amber Baber
Village Gardens
Program Director

This? Or this?

Kids' Snack Corner

Charles Roberton
Village Market
Community Volunteer

Village Market
New Columbia*
Portland, OR
25 August 2011

* Oregon's largest affordable housing community

CAN A CORNER STORE SURVIVE WITHOUT HIGH PROFIT ITEMS LIKE TOBACCO AND ALCOHOL? VILLAGE MARKET MAY HAVE THE ANSWER.

volunteer education and more nutritious food choices (including fresh produce and prepared foods) for the residents of New Columbia. Village Gardens provides for a volunteer...

public transportation, these residents must shop for their food at gas stations and corner stores. Village Gardens

Many people living in public housing don't own cars. When supermarkets are not within walking distance or easily reached by

tahini

Julie

MICRO DISTRIBUTION
Hummingbird supports local farmers
by providing distribution of their
products (in this case organic crops)
and also plays a role in transitioning
conventional growers toward organic
agriculture by providing access to a
growing base of organic consumers.

MICRO
Processing foods
ingredients elim
chain and cen
makes two raw s
chocolate filbert br
almond butter, tah
regional snack b

REGIONAL FOOD HUB

A centrally located facility used for the aggregation, storage,
processing, marketing and distribution of regionally produced food produc

HUMMINGBIRD'S CUSTOMERS DON'T WANT THEIR FOOD FROM CHINA, SO THE COMPANY FOCUSES ON

in wheat, teff and flour, traditional polenta and lentils, local honey, local organic filberts and dried cranberries, organic blueberries and prunes, organic cornmeal and wild rice, organic black beans and flax seed (this is only a partial list)

Food made from locally-sourced
the need for a lengthy supply
a region's food needs. Hummingbird
granolas, sprouted almonds,
cal filbert butter, California
r local cranberries and a
ll organic).

Hummingbird Wholesalers
Republic of Eugene, OR
27 January 2011

They work with 16 regional farmers and carry organic food

Hummingbird distributes 225 different products from San Francisco to Seattle (85% of its customers are in Oregon). They

G FOOD SECURITY FOR THE LOCAL COMMUNITY

FOOD HUBS and THE ART OF COOPERATIVE DISTRIBUTION

I travel to Eugene, Oregon. As often happens, I'm blessed by the able assistance of a well-informed authority on the local food scene. Erin Walkenshaw is snugly wrapped in a coat, scarf, and woolen hat. While it's not raining this January morning, a cold fog blankets everything. I have ideas of things to photograph—the pioneers of organic certification at Oregon Tilth, a seed grower engaged in a genetic-drift lawsuit against a maker of GMO seeds, a **FARMSCAPER** who builds **BEETLE BANKS**—but Walkenshaw has other plans.

"Eugene is all about distribution," she tells me between sips of steaming tea, her thermos careening around her feet as she drives me across town. "You can get farmers anywhere, but not pictures of infrastructure."

We start at Hummingbird, a modest-sized warehouse lined with industrial shelving that nearly reaches the ceiling thirty feet above my head. The story is inside those boxes. Almonds. Filberts. Purple Karma hull-less barley. Pumpkin seeds. Flax. Rye berries. Hard red winter wheat. All grown by local farmers. A **REGIONAL FOOD HUB** like Hummingbird aggregates the output from a hundred farms across the state, puts their product on trucks, and sends it across the Pacific Northwest. In-town deliveries—their biggest market—are made by bicycle (this is Eugene, after all). **Regional food hubs make local food movements a reality; by rebuilding local infrastructure they provide a critical bridge between producer and consumer.**

Walkenshaw and I eat lunch in the car, parked in an alley facing the back of an old brick building. She's brought homemade bread and two mason jars filled with thick potato soup. Assorted bicyclists pass. She mentions a job stocking produce at a local supermarket, her life on a nearby organic farm, and her involvement with the local chapter of **SLOW FOOD**.

The supermarket job is especially enlightening. Walkenshaw explains why local produce never seems to end up at your local supermarket. Since most supermarkets are stocked by trucks coming from a regional center five hundred miles away, local produce

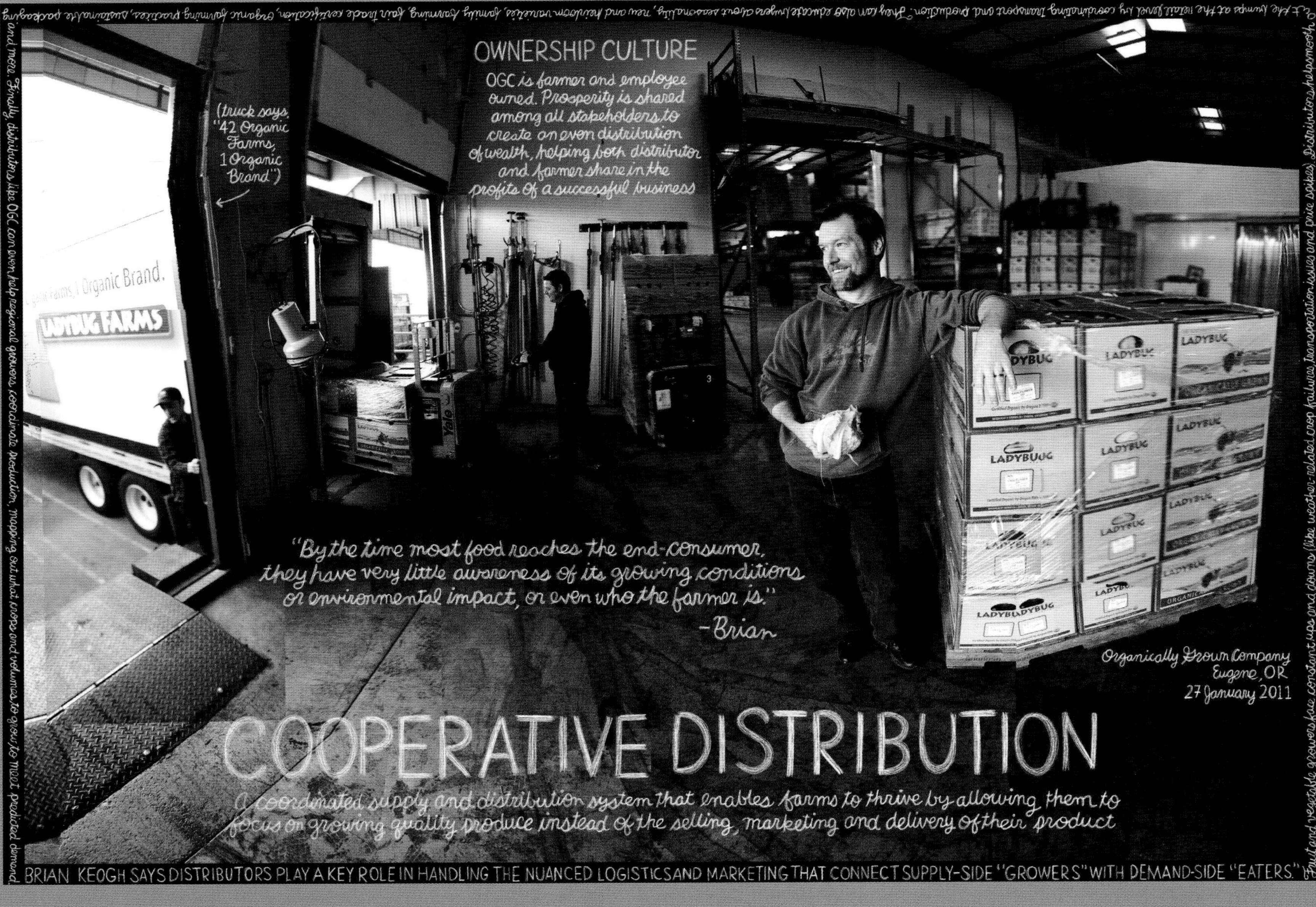

COOPERATIVE DISTRIBUTION

Coordinating supply and distribution system that enables farms to thrive by allowing them to focus on growing quality produce instead of the selling, marketing and delivery of their product

"By the time most food reaches the end-consumer, they have very little awareness of its growing conditions, its environmental impact, or even who the farmer is."
—Brian

Organically Grown Company
Eugene, OR
27 January 2011

OWNERSHIP CULTURE

OGC is farmer and employee owned. Prosperity is shared among all stakeholders to create an even distribution of wealth, helping both distributor and farmer share in the profits of a successful business

(truck says "42 Organic Farms, 1 Organic Brand")

BRIAN KEOGH SAYS DISTRIBUTORS PLAY A KEY ROLE IN HANDLING THE NUANCED LOGISTICS AND MARKETING THAT CONNECT SUPPLY-SIDE "GROWERS" WITH DEMAND-SIDE "EATERS."

...at the farms by coordinating transport and production. They can also educate growers about seasonality, new and heirloom varieties, family farming, fair trade certification, organic farming practices, sustainable packaging and more. Finally, distributors like OGC can even help regional growers coordinate production, mapping out what crops and volumes to grow to meet predicted demand.

Fruit and vegetable growers face constant price concerns and demand (like weather-related crop failures, transportation issues and price spikes). Distributors help smooth...

would have to be sent five hundred miles to the center, repacked on another truck, then sent back. That exercise would have to be repeated for every local farmer selling produce to every supermarket in every different town in a region. It's called "centralized distribution." "I think about this stuff all day long," she tells me. "Logistics. Distribution. It's maddening."

She's right. To visualize what a sustainable food system looks like, imagine a pie chart with equal slices labeled food production, food distribution, and food consumption. Greater sustainability on a local level comes from optimizing how a community deals with each of these slices. **The challenging piece, unsexy because it depends on things like aggregation and logistics, is distribution.** A farmer can grow produce, but without a customer these goods sit in the field, rotting until a tractor tills it back into the earth, where it hopefully provides nutrients for subsequent harvests that will find a market.

The Organically Grown Company (OGC), based in Eugene, is owned by its employees and local farmers. The sprawling complex sits five miles north of the city and features receptionists, lobbies, conference rooms, loading docks, forklifts, and a fleet of brightly painted trucks. They collect produce from organic farms across Oregon and Washington, deliver it to centralized warehouses, process it, then repack it for delivery to grocery stores in seven states. The secret of their $100 million business also includes a bit of trucking logistics, marketing, and accounting. All play a part in **COOPERATIVE DISTRIBUTION**.

OGC doesn't just distribute certified organic produce (it is the *O* in the company's name, after all); it distributes values. The employees organize organic conferences for local growers. They educate produce buyers, chefs, and consumers about seasonality, new and heirloom varieties, family farming, fair trade, organic farming practices, and sustainable packing. And by literally showing up at their doorstep with fresh organic food, they shift the purchasing decisions of big supermarket chains. In business school, OGC would be defined as the classic market disrupter.

Sandi Kronick is a market disrupter as well. An Oberlin College graduate, Kronick ended up ten years ago in Durham, North Carolina, where she helped found Eastern Carolina Organics (ECO), a grower-owned company that helps farmers' markets distribute their organic produce. It's a company that's changed the region's prevailing "middleman system," one where distributors have no personal relationship with their growers.

"When I first moved to North Carolina to work in organic agriculture," she recalls, "I hopped into someone's pickup truck to visit farmers, and had a moment—I think we were loading strawberries—when I said, 'Wow, there are no payment terms on this invoice. Customers aren't going to know when to send the money back. And you guys probably aren't that good at following up and harassing when you don't get paid, either.' That was something we could do and hopefully provide a better livelihood for these farmers."

While ECO is a mission-driven company, Kronick's not afraid to identify her business for what it really is. "About three or four years ago there was almost a collective 'aha!' moment in the sustainable food scene," she says. "'Oh, wow, logistics, trucking.' Suddenly all these technical things that were really unsexy in the public many years ago are quite sexy now. That's really where the answer is: logistics. **We need to not be afraid to just focus on the logistics aspect of it; the rest can come afterward.** That ability has helped really develop the **FOOD HUB** movement, because people are not afraid to dig into the more boring details of logistics, trucking, and food handling, rather than the more exciting stuff like food demos."

This is the new Marshall Plan. It's how infrastructure is rebuilt. Building personal relationships, thinking small instead of big, and following the principle of short supply chains—reducing the distance food has to travel—strengthens the connective tissue between all players in a local economy.

Pie Lab was founded in 2008 as a pop-up cafe, design with a single purpose: to bring life back onto Gree main street ... by serving pie.

miss Deborah the manager

miss may the customer

PIE = COMM

Pie + ideas = Conversation
Conversation + Design = Social Ch
(pie is easy to linger over and conversations happe

ll about the movie theatre, the stores and the pool hall, he and his buddies would hang out in. His description was one of an exciting place to be. Then he looked around PieLab, at the new businesses on the street, and said, "it's finally coming back to life."

I'M A BIG FAN OF JAMES HOWARD KUNSTLER'S "GEOGRAPHY OF NOWHERE", SO I ASK MISS DEBORAH HOW PIE LAB SHOWS THAT WE CAN BRI

Pie Lab is a partnership between "Hale Empowerment & Revitalization Organization" (HERO) and a design collective known as Project M.

...ivic clubhouse Alabama's

andy the local shopkeeper

NITY

n people linger)

Pie Lab
Greensboro, AL
25 June 2011

...CK TO THE MAINSTREETS OF OUR SMALL TOWNS AND SHE SAYS,

THE NEW MAIN STREET, or
PIE = COMMUNITY

If you take a map and draw lines leading away from Tuscaloosa and Montgomery, Alabama, and Meridian, Mississippi, they'll intersect at a small town with a mostly boarded-up main street, Greensboro, Alabama. Even though it's the self-proclaimed catfish capital of the state, you'd probably drive past it, which almost happened to me. I was in a hurry—I'd risen before dawn to photograph various examples of **BYCATCH** with Biloxi shrimpers and needed to make a dinner that same night at Birmingham's Highlands Bar and Grill—yet I still made a hefty detour, bypassing Highway 65 in favor of country roads much less traveled, all because of an impulsive decision to check out a place called PieLab.

A group of designers led by John Bielenberg wanted to know if good design could actually change the way people live. **Their theory was that people would be more likely to share ideas that improve their community over coffee and a slice of pie in an inviting locale.** Bielenberg's Project M partnered with H.E.R.O. Housing, a local nonprofit focused on rural development, and opened the first new business this town's main street had seen in decades. **PieLab is a social experiment that plays out in a public space. It shows how food helps build communities.**

Their enterprise has a number of moving parts. First, PieLab itself is a coffee shop, a community center, a computer lab, and a large performance space. Second is Pecans!, which mentors local kids in a pecan brittle business, with its proceeds going to support a student scholarship program. Third is a bamboo bicycle–making studio. Who knew Alabama had so much bamboo, let alone knew what to do with it?

Can a cleverly architected coffee shop bring a shuttered main street back to life? I discuss this over—what else?—a slice of pie with Miss Deborah, PieLab's manager. She explains that they didn't create PieLab to replace what was missing from the community. They just wanted to bring their community closer together.

I sometimes ask people for their definition of sustainability. Quite often they don't have one. They

67

Melvin

Alveter

PECANS!

LOCAL STUDENTS
LOCAL INGREDIENTS
ENDLESS POSSIBILITY

made in Greensboro, Alabama

YOUTHBUILD
scholarship fund

(a secret recipe)

(local pecans)

"The PECANS! Project keeps turning a cycle of learning. This helps sustain the community over time because youth ARE the community; when we learn everyone benefits."
— Alveter

YOUTHBUILD
A national non-profit that serves out-of-school, at-risk youth through job-training, GED courses and leadership development. YouthBuild in Greensboro works with 52 youth a year.

Pie Lab
Greensboro, AL
25 June 2011

Profits from Pecans! are deposited into a fund which distributes three $500 scholarships to students each year to pursue a post-secondary education.

LOCALIST

Using a local resource (students + pecans) to create a small business (and college scholarships)

THE PECANS! PROJECT IS A SMALL-BUSINESS DEVELOPED BY YOUTHBUILD* STUDENTS IN 2009 TO CELEBRATE A LOCAL RESOURCE AND CREATE A SELLABLE PRODUCT.

claim that ad agencies reduced the term to meaningless marketing-speak long ago. They're right. **But *sustainability* is an important word, one worth fighting for. One worth taking back.** When people do offer me a definition of sustainability, it's always interesting because it inevitably represents how they want to live.

Miss Deborah's definition of sustainability doesn't come that afternoon. Instead she gives it some thought. Her answer arrives by email months later.

"We use lumber from a torn-down barn to make something new," she writes. "We recycle items that have been discarded. We source our food from local farmers and local organic companies. We offer job training, promote entrepreneurship, use only eco-friendly products, offer leadership classes that teach responsibility to the community, and teach young people to build, plant, and care for gardens that are then given to the community. Sustainability doesn't just happen. It takes action!"

These changes in Greensboro started with a slice of pie, food being a most efficient social lubricant. And in case you're wondering, Miss Deborah's favorite pie is apple—with a scoop of vanilla ice cream. The recipe? Why, she was afraid you'd never ask.

MISS DEBORAH'S PIELAB APPLE PIE

Yield: One 9-inch pie

INGREDIENTS:

5 cups peeled and sliced Granny Smith apples
3 tablespoons lemon juice
1 tablespoon grated fresh ginger
¼ cup flour
¼ cup granulated sugar
¼ cup brown sugar
3 teaspoons ground cinnamon
1 teaspoon ground nutmeg
2 portions pie dough; use favorite recipe
1 egg white mixed with 1 drop water
Turbinado sugar, for sprinkling

DIRECTIONS:

1. Preheat oven to 350°F.
2. In large bowl, mix apple slices with lemon juice and ginger.
3. In a separate medium bowl mix flour, sugars, cinnamon, and nutmeg.
4. Add flour mixture to apples, gently tossing apples until thoroughly coated.
5. Roll out two pieces of dough; transfer one piece to 9-inch pie plate. Pour in apple filling.
6. Make a lattice top with the second piece of dough.
7. Lightly brush egg white mixture on top and sprinkle with Turbinado sugar.
8. Bake for about 45 minutes, until bubbling.

NEARLY LOCAL: THE STORY OF CONNECTED MARKETS

Silicon Valley is a highly competitive marketplace, not just for technology but also for employees. To attract the top talent, companies offer an extraordinary range of perks that include free transportation, massages, dry cleaning, day care, sleep pods, full-service gymnasiums, movie theaters, and meals. Google outperks everyone. It employs a roster of seriously overqualified chefs, each entrusted with a conceptually themed restaurant that serves employees three squares a day, for free. Dan Watts is at the center of this culinary maelstrom. As head of purchasing he seeks out the finest fig growers, the most sustainable oyster beds, the plumpest blackberries, and the freshest line-caught fish, then buys as much as he possibly can. **Google serves thirty-five thousand meals each day. To put that in perspective: Many San Francisco restaurants don't serve that many in a year.**

Watts and the Google culinary team ask me to document some of their more exotic producers, which include the fishermen of Yakutat, Alaska, who work closely with the state's Department of Fish and Game to sustainably fish salmon on the Situk River. A week later I join Shannon and David Negus, two gill-net salmon fishermen who venture up from Oregon to Yakutat each June for the area's hectic, adrenaline-fueled four-month-long fishing season. Yakutat is less a town than a village. Less a village than a random collection of half-completed plywood-lined structures built more for surviving Alaskan winters than for any potential suburban curbside appeal. It's a place with no bed-and-breakfasts. No supermarkets. No restaurants. No dog walkers on Main Street. In fact, for the next four days, I eat all my meals at the kitchen table of the Neguses' sprawling semi-constructed two-story home, which is a good thing, because Shannon really knows how to cook.

The dinner conversation with this couple is easy. Shannon is clear-eyed, earnest, and quiet. David is well-educated. A formulator of theories. Various texts scattered around the house leave the impression that he could have been a professor of marine sciences (it turns out he almost was). The two often finish each other's sentences or tell me what the other is thinking. They clearly spend a lot of alone time together.

I'm put up at a hunter's vacant cabin just down the road where I'm advised to remain watchful for bears—true enough, I find scat within minutes right outside my door. I'm also told to get some sleep because the following day promises to be long.

This is an understatement.

We start before 6 a.m. I would say dawn but it never gets truly dark at this time of year in Yakutat, more like dirty gray. As for how gill-net fishing works, it's fairly straightforward. A net 120 feet long is secured at one end to the shoreline, with the other end wrapped around the skiff's bow. The boat is small, fully exposed to the elements, with a single outboard motor. As we slowly pull away from the shore, the net tightens. Salmon coming in with the tide immediately hit the net, their rhythm staccato, with such force that water sprays everywhere. The air fills with a fine mist. Shannon and David slowly pull the net into the boat, edging closer to the shoreline as they methodically disentangle each salmon. The humpies and chum, both lower-grade species, are gently tossed back into the water. They are commodity fish, primarily of value to canners. Shannon and David are instead focused on sockeyes and the occasional king, two fish with the greatest demand among chefs in San Francisco's Bay Area, where Mystic Salmon does most of its sales.

The Yakutat District set-net commercial fishery offers only 163 permits, and they're awarded for life. These prized objects pass from generation to generation, mostly among Native American Tlingit—their name roughly translated means "People of the Tides." On rare instances permits are sold, which is how Shannon and David got theirs. That event led to the birth of Mystic Salmon.

During fishing season the river is lined with boats. Some of the fishermen even spend the summer season in cabins dotting the shoreline. It's a closed group with few visitors. While many fishermen here immediately toss their catch

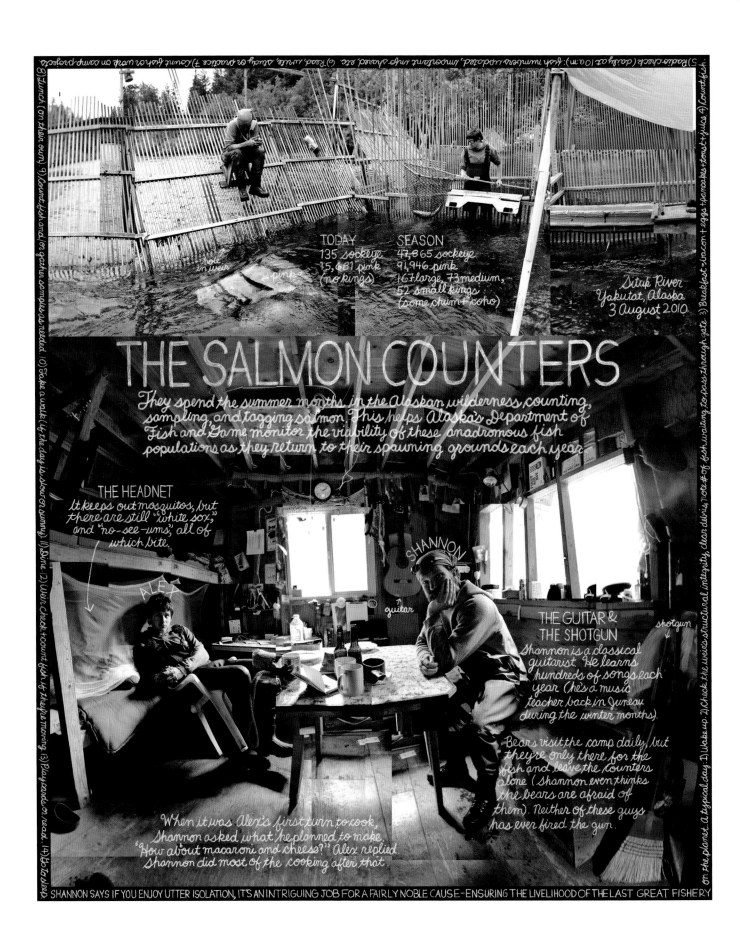

into massive ice chests as they work, David meticulously bleeds out and ices each fish as it comes from his net. This process takes up valuable time out on the water, but it delays rigor mortis and bacteria growth while extending the salmon's shelf life. **These aren't things you would immediately see when buying your cut of salmon at the store, but David knows, so he takes the time to do it.** The salmon end up in San Francisco less than thirty-six hours later.

When we hit the shore a waiting truck takes them ten miles inland, to a structure David calls his "processing plant," though you and I would consider it a refrigerated mobile home with an ingenious icemaker built into the far wall. The fish are gutted and cleaned by Shannon and David's eldest son, then packed with ice into cardboard boxes. A clipboard on the wall makes note of chefs—all referred to by first names—and final destinations. If you live in San Francisco, you'd undoubtedly know the restaurants. Once boxed, the fish is dropped at the airport. Yes, the airport. Yakutat might not have any restaurants or grocery stores, but it does have an airport with daily Alaska Air flights. Eight hours later cartons of Mystic Salmon sockeye and king land at San Francisco International Airport.

The next morning I set out in a light rain to see the weir farther up the Situk River. My guide is Gordie Woods, the wiry and wind-burned area manager for Alaska Fish and Game. We meet on a gravel beach strewn with dozens of gnarled, waterlogged tree carcasses, their roots splayed like distended ebony fingers. These trees seem ominous, and are clearly a problem, as I learn after another Shannon, Shannon Spring, joins us. Spring has descended from the weir and now returns with us upstream, helming an open-hulled aluminum skiff around trees that clog the waterway. At times we bang against them, but it's worse when they're submerged only inches below the river's surface; they can literally rip a boat apart, sending everything overboard, so Woods mans the bow, vigilantly eyeing the water. The engine is loud, too loud for conversation, so I hang on to my camera gear and simply stare at the tree-lined banks rising sharply on either side of us. This place is beautiful.

We arrive at the weir an hour later. It's a Mad Max contraption. Tentlike structures framed with PVC pipe rise nonsensically out of the stream (I later learn they're actually a precisely designed mobile research lab for studying steelhead salmon). A muddy path leads away from a makeshift dock, disappearing into the trees. Finally, we reach the weir itself. This modest and somewhat underwhelming construction is a rickety fence nearly fifteen feet tall composed of aluminum conduit pipes (called pickets), each spaced three-quarters of an inch apart to allow water and small fish passage, while keeping adult fish—like the sockeye—from passing through. Pulling up five or six pickets creates a gate, which the fish immediately find to continue their journey upstream. Right now the gate is closed and thousands of salmon loiter, their dorsal fins tracing faint anxious curves as they crest the water's surface, darting left then right then left again as they search for a way through.

Shannon Spring sits above this opening for twelve hours each day, counting salmon with a fancy handheld device. It has four different number clickers: one for each pink, sockeye, silver, and king. This data is radioed down to Woods at the base of the Situk, entered into a computer database, then compared with data from previous years. These numbers eventually determine how many fish Shannon and David Negus take out of the water at the Situk's terminus and send to restaurants in San Francisco.

Spring invites us up to his cabin for coffee. We climb the steep mud path behind the dock and arrive at a raw timber and plywood cabin with a single twelve-by-twelve-foot room. The inside is sparse but carefully organized, with a wooden slab set to one side for cooking, a table, and a bunk bed. An outhouse and outdoor shower stand across the clearing. Spring's monastic forest life lasts for the entire five-month season. He sometimes has assistants, but mainly what he sees are bears. They torment him each day, and who could blame them, with thousands of salmon trapped below the weir, waiting to pass through the gate? They're easy prey, and though Spring keeps a can filled with large rocks beside him to scare the bears off, the sandy beach is still littered with hundreds of half-eaten fish.

nt escapement onto spawning grounds, and manage time, area and gear types to attain escapement goals.

necessary to perpetuate the species while allowing its optimal harvest. Yakutat Fishermen rely on daily escapement information from this weir.

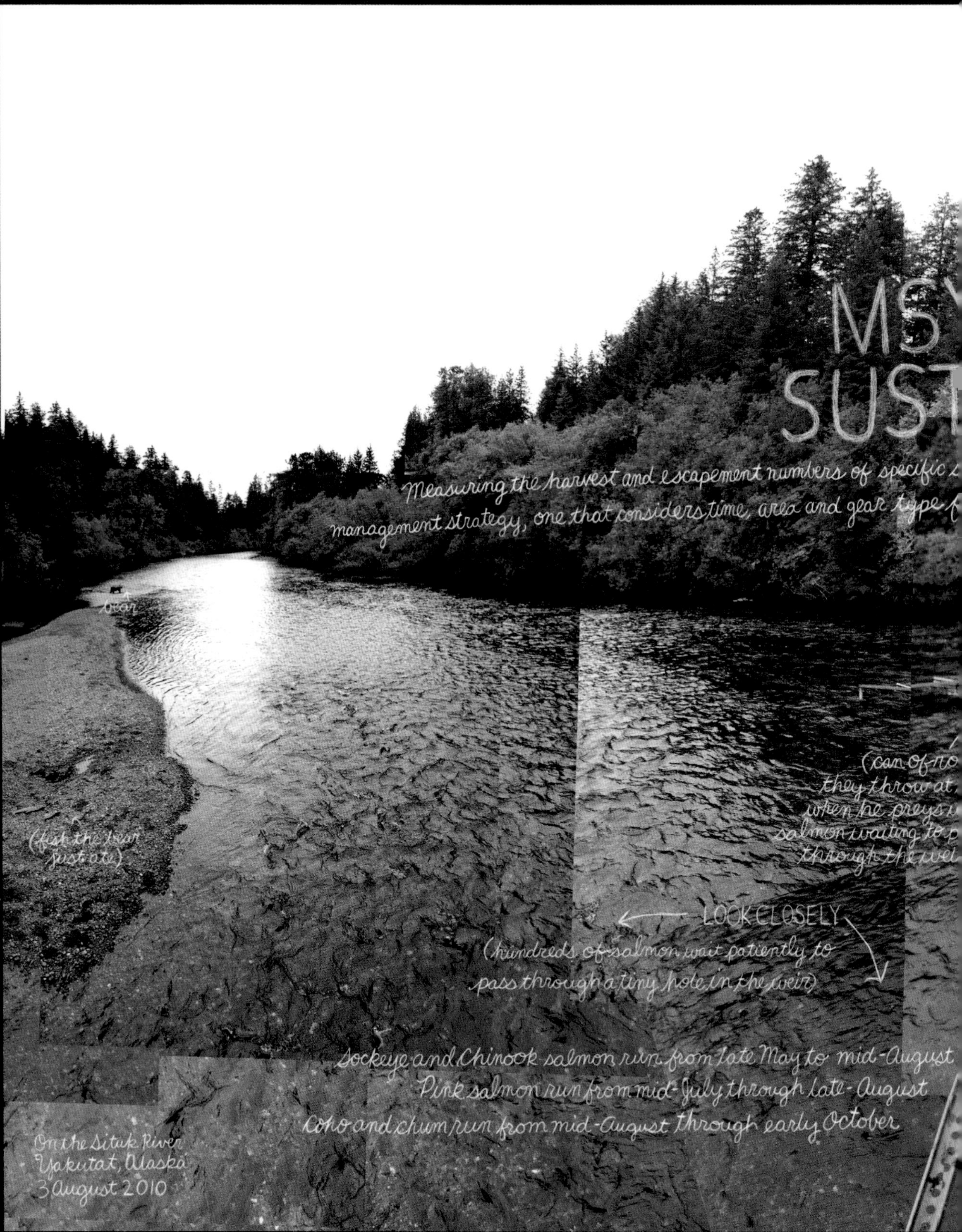

MSY
SUST
Measuring the harvest and escapement numbers of specific
management strategy, one that considers time, area and gear type
bear
(fish the bear just ate)
(can of ro
they throw at
when he preys
salmon waiting to p
through the wei
LOOK CLOSELY
(hundreds of salmon wait patiently to pass through a tiny hole in the weir)
Sockeye and Chinook salmon run from late May to mid-August
Pink salmon run from mid-July through late-August
Coho and chum run from mid-August through early October
On the Situk River
Yakutat, Alaska
3 August 2010

TO DEFINE MSY ASSUMES A HIGH LEVEL OF UNDERSTANDING OF THE TARGET POPULATION (BIOMASS, L

MAXIMUM
INABLE YIELD

assists   Alaska's Department of Fish and Game in their deployment of a viable fishery
hing   different fish. The goal? Provide for optimal harvests while sustaining each salmon species

ALEX

the "trap"

← the counter

SHANNON

Each morning Sockeye and Chinook salmon are
sampled in the "trap". Research includes:
1. Mark / Recap projects on returning adult salmon
2. DNA analysis is in every system in Yakutat
3. Age, sex and length sampling

GORDIE

sockeye

WHAT WENT THROUGH THIS WEIR IN 2010
167 large, 73 medium, 52 small Chinook
47,865 sockeye
91,946 pink salmon

LE, AGE GROUPS + HABITAT UTILIZATION). MOST WORLD FISHERIES DON'T HAVE THIS INFORMATION AVAILABLE.

While Spring makes coffee, Gordie Woods explains his scientific justification for all this diligent activity associated with the weir: It's the principle of **MAXIMUM SUSTAINABLE YIELDS**, the goal being to collect enough measurable data to guide a bountiful harvest while still ensuring the viability of a species. Part of his explanation eludes me, though I do hear rough calculations that for every 3.2 sockeye passing through this weir, one can be safely fished out at the terminus of the Situk. Woods and Spring are reluctant to draw conclusions. The data isn't precise enough yet. Plus, so many factors impact on a sockeye's life, from ocean predation to genetic instability and now—with climate change—unpredictable fluctuations in ocean temperatures. If the water is too warm, the salmon won't spawn.

The hard work and perseverance of a few men at Alaska's Department of Fish and Game, spanning decades, provide a window into how these salmon live, at least on one river, and how understanding the principle of maximum sustainable yields might lead us to actually safeguard the viability of salmon in this corner of the world. I came to Yakutat to document a more sustainable, accountable way to harvest a precious natural resource, and now I have it. **Fishermen on the Situk River's terminus work under catch limits set by a warden straddling a weir two hours upstream.** The image I create is called "Terminal Fishery," at least for the next few hours.

Right before boarding my plane, Shannon Negus pulls me aside, and with uncharacteristic frankness says I absolutely cannot call the image we've created "Terminal Fishery." She knows it's why I came, even why I find it so important, but she thinks it's merely a subset within a much larger concept. **Many consumers are now fixated on all things local, but no region can provide you with everything.** What about coffee or sugar, or even salmon? You don't have to live in Guatemala just because you like bananas.

"You need to explain the principle of **CONNECTED MARKETS**," she intones as they announce the final boarding. "Explain the idea that the same principles you use to buy things from within your community should be applied to what you buy from halfway around the world. Just because you can't see where something comes from, it doesn't mean you don't need to know how it was made."

When you buy salmon, you have options. It can be fresh or previously frozen. Farmed or wild. Pink, sockeye, or king. But do you know *how* the fish was caught, *when* it was caught, *where* it was caught, and by *whom*? **The industrial food system is both efficient and opaque. Half the time we have no idea what we're eating, especially when rolling down the center aisles lined with shelves of processed foods at the grocery store, their ingredient lists only decipherable by someone with a graduate degree in food chemistry.**

Shannon Negus's idea of the connected market reframes the concept of choice. A local food system might provide everything you *need*, but not everything you *want*. It also reframes the concept of **LOCAL**. Food is made by people who work for companies—both big and small. The choices these people make—the ingredients they use, their production methods, their labor practices—define their own value systems. When you buy food, you're buying these values. Their values. Becoming an informed consumer means understanding first what you hold dear, what you cherish for you and your family, then finding the products that align with those principles. If they're local, fine. If they're not, that's fine, too.

Values trump distance.

To live in a connected market requires the physical act of becoming not just educated but involved; buying salmon from Shannon and David Negus—either from a restaurant that serves their fish or by ordering your own directly from their website—affirms your belief in their values, in how they fish. It also confers your approval of the values behind Situk River fishery management practices. It's an entire story, one that rewards the consumer for asking questions . . . and finding the right answers.

The connected market principle isn't a certification. It has no third-party verifiers. What it offers is personal stories based on relationships. Still, I wonder what a connected market certification would look like. What values would Shannon and David Negus share with consumers who understand the limits of their local food systems?

CONNECTED MARKETS

When producers and consumers can envision each other — even across great distances — a product (like salmon) transforms from a commodity into a carefully guarded, precious resource

2 August 2010
around half Nakuka
Situk/Lost River
Yakutat, Alaska

DAVID   A connected market is a mindful market.

My wife's goat's milk ice-cream company reaches a crossroads fairly early on. First, Laura appears on the *Today* show. Articles pop up in *Newsweek,* the *New York Times,* and the *Wall Street Journal.* Everyone writes about the lady making the only goat's milk ice cream in the country. Laura's motto—which she repeats like a mantra—proves true: "Who doesn't love ice cream?"

Still, industry veterans take a more sober view. One comments that her potential market is "a niche of a niche." He might be right. The product is expensive and unusual. If she stays small, continues to use Petaluma goat's milk and costly ingredients, and sells exclusively at local farmers' markets and specialty stores, Laloo's

may never find a large enough audience to reach profitability. The financial stress builds. In gambling terms, we're all in. While it mostly remains unspoken, knowing the grim facts keeps us awake at night.

The company slowly grows, expanding from one employee to seven. The "office" moves out of a spare bedroom and into a freshly remodeled garage, then finally gets its own barn. An official phone system is installed. People start wearing Laloo's T-shirts. Children mail in drawings of goats. A board of directors assembles. Those first meetings are fractious, as any business built on passion that now receives outside counsel from return-conscious investors tends to be. Whole Foods starts distributing the product, first in

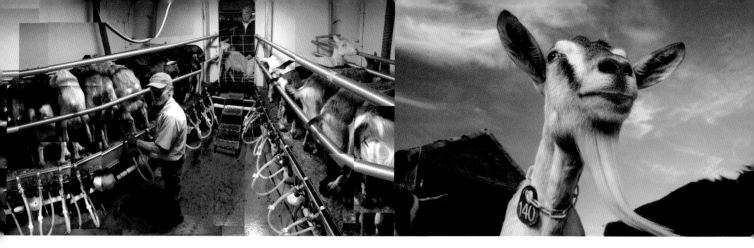

Northern California, then the entire state. By the end of the year it's sold across the United States. The highest sales come from an island only thirty-four square miles in size: Manhattan.

To reduce **FOOD MILES** to the East Coast, Laura resolves to add a production facility in the Midwest. She makes a series of midwinter trips to visit Wisconsin goat farmers and eventually helps them create a goat's milk co-op. She assists in drafting their marketing materials, finds buyers for excess milk she can't use, and even has me shoot promotional photographs of the goats "in action." Lastly, she makes a deal with a Wisconsin co-packer and starts ice-cream production.

The transformation is sudden, especially after Laura's "lactose-friendly" message finally connects with a large consumer base: lactose-intolerant shoppers unable to eat ice cream from cow's milk. For many, Laloo's becomes the first ice cream they've had in years. They flood Laura with emails offering tearful thanks.

Laloo's now knows its buyers. They aren't just from Petaluma. They're from across America and equally split between trendy-savvy foodies, the nutrition-conscious, and those with dietary concerns. This also means that Laloo's is no longer a local product. It caters to a community of no fixed geography, yet one that's mindful of what it eats. Laura listens to them, then responds by transforming her brand message to address more national concerns. She creates a crew of Laloonatics around the country, exuberant, mostly college-aged kids who start doing outreach and demos. She starts a product tie-in with Waterkeeper Alliance, raising funds for Robert Kennedy Jr.'s nonprofit to help family farms clean up their waterways. And she becomes one of the

first food companies awarded **B-CORP** status, a certification for companies with sustainable business practices. At the same time she's executive director of the Lexicon of Sustainability. She also sits on the board of our town's community farm, which provides food for a local health center, works in our daughter's school garden, and still gives away ice cream to nearly every fund-raiser that asks. She's local and nearly local, just as things should be in a connected market.

ORGANIC FLOWERS

No use of artificial or chemically derived nutrients, pesticides or herbicides. Minimal use of organic approved herbicides and pesticides.

Joel

Lyle is the accidental farmer

↑ (black-eyed susans)

THE ACCIDENTAL FARMER
Lyle was raised in NY's Hudson Valley. His father subscribed to Organic Garden magazine (this was back in the 50's). When he bought this land he didn't intend to start a market farm. It just happened. Now they grow over a hundred varieties of flowers and vegetables on 35 acres.

Pastures of Plenty
Boulder, CO
8 August 2010

I ASK LYLE WHY FLOWERS SHOULD BE ORGANICALLY GROWN IF THEY AREN'T GOING TO BE EATEN.

 jected to some of the worst human rights violations, not to mention environmental pollution, hence his decision to go organic.

and pesticides) used in conventional agriculture. Because flowers aren't eaten, workers in this industry are sub-

He says that conventional flower production requires the same chemical inputs (herbicides, fungicides,

Your local supermarket sells more stories than your local bookstore. And just like your bookstore, these stories—an artful mix of fact and fiction—are placed on shelves. In the supermarket, these stories are told with pictures featuring pastoral images of barns, grain elevators, and livestock leisurely grazing on impossibly green meadows. Turn the box over and you'll find that the story continues with descriptions using the words *natural*, *wholesome*, and even *sustainable*.

When you buy food, you're buying values. You're buying the values of the company that prepares and packages your food. The values of the company that transports it. And the values of the company that sells it to you.

As a consumer, your values are tempered by economics. You can only buy what you can afford. Still, some consumers are uncompromising. They demand organic food and search for the USDA's identifying symbol.

Organic certification doesn't tell you everything you need to know. It doesn't tell you about the working conditions of farm laborers, if growers are paid a fair price for their goods, or even if these growers are responsible stewards of their lands, ensuring their use by future generations.

But it does hint at a product's "story." Consumers can pick up a product, study the packaging, search for the identifying symbols representing hundreds of certifying agencies, and connect the values of those companies with their own purchases. This is how a connected marketplace works. Far from a theory, its manifestations already appear in a variety of forms, with the most obvious being **ORGANIC**, **FAIR TRADE**, and **DIRECT TRADE**.

ORGANIC

Warren Weber has an **ORGANIC** farm in Bolinas, California. If you know where Bolinas is then you know Warren. The town is on a map but no road signs will direct you there. Whenever a sign goes up, the locals yank it down. It's a town apart. People there are just different, Warren included. His hundred acres are the oldest continuously certified organic row crop farm in California.

I ask Warren about the good old days. He claims the best thing that happened to small farmers was industrial agriculture. It forced them to do something different to survive in the marketplace. To him, doing "something different" meant going organic.

While there are many types of agriculture, consumers mainly see two: organic and conventional. Organic farming uses natural inputs that enhance **SOIL FERTILITY**. That means nothing is used that might prove harmful to the air, the water, or the soil. Conventional farming uses petrochemical-based herbicides and fertilizers. Their use has been linked to water and air pollution and soil decontamination. Consumers concerned about the **EXTERNAL COSTS** associated with conventional agriculture—things that may affect their health and the environment—often buy organic products.

The USDA has designated organic certifications for meat and dairy products, produce, flowers, and even clothing made from cotton.

The most personal question you'll ever ask small farmers is whether they're certified organic. Some will heartily reply, "Yes!" Others may consider themselves organic in practice but have opted out of the federal government's organic certification program. This means they can't call themselves **CERTIFIED ORGANIC**.

As Eliot Coleman, a farmer and author of *The New Organic Grower*, points out, "I've never been organically certified because I don't believe we know enough to say exactly what practices create the most nutritious food. That's been taken over by the USDA. They stated when they first rolled out organic certification that they didn't believe it was any better than other food, so it's a little hard to join up and play

their game. We try to be law abiding. We don't call our produce 'organic,' though we do insist on calling this an 'organic farm.'"

To put this schism in context, Coleman takes us back a few years. "In the United States, the left had taken over organic farming, but it started in England with the right wing. A lot of the early organic pioneers were 'conserve-atives'—old-time Conservatives." These defenders of England's rich agriculture traditions included Sir Albert Howard, author of *An Agricultural Testament*, who inspired Lady Eve Balfour and Alice Debenham to conduct the first field trials comparing organic and chemical-based husbandry. The "Haughley Experiment" looked closely at the entire food chain, from soil to plant to animal. Thousands of tests conducted over two decades in Haughley Green provided consistent and conclusive evidence attesting to the benefits of organic farming. Looking back at those times, Eliot Coleman ruefully concludes, "You wonder . . . maybe conservatism changed and today's supposed conservatives really aren't into conserving."

Back in the United States, farmers were inspired by *Organic Farming and Gardening,* a magazine first published by J. I. Rodale, while the "back to the land" movement later led tilth organizations in the Pacific Northwest to pool their collective organic farming knowledge. The Willamette Valley Tilth chapter eventually offered organic certification to its own members. I ask Harry MacCormack, one of the members, how that happened.

"We all knew and sold food to each other but then Bob Cooperrider got to selling to people out of state," he explains. "Bob needed something to explain what he was doing, to prove he had it right. He actually wrote the first standards—they were one page—and there were twelve of us that certified one another with that one-page standard. My humble farm, where I'm sitting right now, was one of the first farms certified under the Willamette Valley Tilth standards."

I track Bob Cooperrider down a few weeks later. He now travels the world advising farmers on sustainable practices.

Warren has the oldest continuosly certified organic farm in California.

WHY GROW ORGANIC? Apart from not polluting the soil, the aquifers, and ourselves, growing organic ensures that you are building soils for future generations and doing so in a way that maintains a balance of critical natural resources.

Star Route Farms, Bolinas, CA 16 December 2010

WHY BUY ORGANIC?
1. health of the soil
2. health of the plants
3. health of the workers
4. health of the consumers

ORGANIC
Farming using natural systems and inputs with a view toward a sustainable future.

WARREN WEBER SAYS THAT 35 YEARS AGO, MOST "EXPERTS" THOUGHT ORGANICS COULDN'T PRODUCE

"So what happened to that piece of paper?" I ask.

Bob gives it some thought, then replies, "I have no idea. I must've thrown that out long ago."

"Why?" I'm incredulous.

"How did I know this whole organic thing was going to get so big?" he answers.

Organic farming is now a multibillion-dollar industry, one that's big enough to involve very big players. The largest organic produce retailer in the United States is Walmart.

If the organics industry has a watchdog, it's Mark Kastel, cofounder of Wisconsin's Cornucopia Institute. He sees an organic-certification system that is now hopelessly bifurcated.

"There are salt-of-the-earth farmers that earn their reputation by being organic. On the opposite spectrum, we have some factory farms milking nine thousand cows organically," he points out. "We've got one hundred thousand birds in a building laying eggs that are labeled organic, birds that never go outside." The whole idea of large organic-certified companies leaves Kastel with more questions than answers. "You can grow a lot of vegetables, but are you really concentrating on soil fertility and the health of your plants? Are you merely substituting chemicals with fertilizer from some organic source that's not from your farm? Are you following a regenerative, sustainable program?"

If Kastel's principles of "true organic production" aren't practical from a production standpoint, and corporate interests have hijacked the current system of certification, we have a problem.

Michael Sligh served as founding chair of the National Organics Standards Board. He's also a farmer. "The core debate we still wrestle with is how to grow the organic paradigm," Sligh explains. "My approach is diverse farms embedded in local ecological systems to match and meet local consumption. Others say, 'Let's lower standards and bring in larger acres faster.' That creative tension remains and becomes more diverse every day because people come to organic now for different reasons. **In earlier times, before there was much of a marketplace, people came for personal or philosophical**

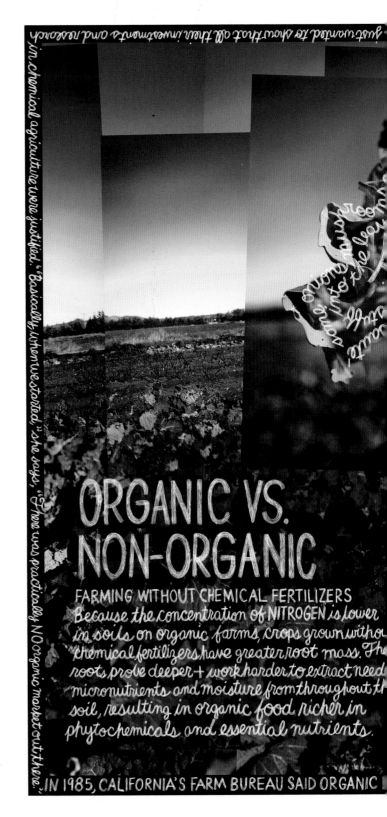

ORGANIC VS. NON-ORGANIC

FARMING WITHOUT CHEMICAL FERTILIZERS
Because the concentration of NITROGEN is lower in soils on organic farms, crops grown without chemical fertilizers have greater root mass. The roots probe deeper + work harder to extract needed micronutrients and moisture from throughout the soil, resulting in organic food richer in phytochemicals and essential nutrients.

IN 1985, CALIFORNIA'S FARM BUREAU SAID ORGANIC

Full Belly Farm
Capay Valley (Guinda), CA
27 January 2012

"We believe in it.
We are just trying
to find balance
between economic
viability and the
kind of farming
that we want to do."
— Judith

...G WAS A FOOL'S ERRAND, A CERTAIN PATH TO POVERTY.   BUT JUDITH AND HER PARTNERS HAD A DIFFERENT IDEA.

Foxglove Farm
Salt Spring Island, BC
26 September 2010

Michael

"The words we use define who we are. 'Organic' was the word some of us have been using for 30 or 40 years—to identify a broad set of social, ecological, and spiritual principles about our farms and how we produced food for our communities. Now the USDA has given the word 'Organic' a legal definition, in essence taking ownership of the word, and limiting its use to a narrow set of rules and regulations designed to support a distribution and marketing system. For some of us the word no longer addresses the deeper issues that were at the heart of the origins of the movement."

"Certification and label systems are like locks on doors, they are there to remind us of our boundaries."

"We've got to find new ways to talk about what we do, we may have to use different words."

BEYOND ORGANIC

There is a fundamental difference between the organic movement and the more recent organic industry. We need to dig deeper and look beyond narrow legal definitions to find a philosophy that truly addresses a system of agriculture that is incredibly complex and multidimensional.

MICHAEL ABLEMAN BELIEVES THAT IN THE FUTURE, FULL-TIME FARMERS SHOULD NO LONGER GROW FRUITS AND VEGETABLES.

"We can all live without another carrot or tomato, but we can't live without pristine sources, and given our resources, these will have to be plant-based.

In community garden plots. There has been entirely too much energy and focus in the food movement on growing that which we could actually survive without. Instead, this should be the responsibility of individuals and families to do for themselves in their front and backyards, on their balconies and rooftops, and

reasons. Now people come for the economic opportunity. These companies are hedging their bets, putting some of their eggs in the organic basket, and they have influence."

"You can do organic on a very large scale," he continues. "You can go out to the upper Midwest and see five thousand acres of diversified organic crops and livestock. They follow the principles and do very well. We have to think about scale within the context of local ECOSYSTEMS and the amount of land necessary to make a living."

I find one observation by Sligh to be particularly interesting. I ask if there was anything he wished the first USDA organic certification had contained, and he replies, "We weren't so much interested in just protecting the earthworms and the birds and the bees and the animals. It was also about protecting the people, the stewards of the land and those that prepare and bring food to the table. It has always been a holistic approach. The institutional phase of defining organics was a bit like driving a square peg into a round hole. Some of our values were not institutionalized in the U.S. certification. One of those values was the social value, the value of fairness for farmers, for workers, fairness for the farm system. There's a growing discussion across the planet. What is fairness in the food system? I think the sweet spot is where you have the marketplace, government support, and consumer demand all meeting at the same time."

Consumers' role is to make purchasing decisions and support a food system that reflects their values.

Since J. I. Rodale's passing in 1971, subsequent generations of Rodales have continued to promote organic agriculture through publications, conferences, and ongoing research. In fact, many consider them the proverbial First Family of American organic agriculture. I've known Anthony Rodale and his wife, Florence, for a few years. We've broken bread together. When I ask about the organic movement's tumultuous history, Anthony takes the long view. "The organic idea is still evolving," he tells me. "It blooms, like a flower. Dies. Then blooms again. Each generation takes it a little further. Some like to 'home-grow' it. When they feel it's become too institutionalized, they walk away. Others like it to be local. That's fine, too. We need the whole world to be organic."

"An ECOSYSTEM is the sum of all the parts of a farm or environment that interact to form the whole. These include the natural (soils, water, sun), the biotic (plants, animals, microorganisms, people), and the social (communities, workers, family). Each part contributes to the whole, but not in a vacuum. Since we have a very poor understanding of how things are linked together (or at least a very superficial understanding), I believe it's better to keep our hands off the ecosystem as much as we can and let those interactions just happen. We should only interfere when we absolutely have to, and when we do interfere, it should be in a way that minimizes the potential side effects of that interference."
—Steve Ela, Ela Family Farms

FAIR TRADE VS. DIRECT TRADE

In the 1980s a group of coffee farmers in Oaxaca, Mexico, organized the Union of Indigenous Communities of the Isthmus Region to get better prices for their coffee crops. Their struggles led them to enlist the support of Solidaridad, a Dutch NGO. The result was the first **FAIR TRADE**-designated certification for a food product.

Fair trade is a now a global labeling initiative. You can find its labels on a variety of products at your local grocery store, from chocolate to bananas, coffee to tea. As Colette Cosner, executive director of the Domestic Fair Trade Association, points out, "Fair trade was originally conceived as a way to address disparities between conditions of small-scale farmers in developing countries from Africa, Latin America, and Asia, collectively referred to as the Global South, and those of subsidized farms and industrial countries in the Global North that have greater access to things like financing, crop insurance, and other advantages.

"The goal," she continues, "was to help farmers in the Global South stay on their land, build stronger rural communities, and not be forced to work on the plantations of big food corporations. Fair trade pioneers agreed that one of the best ways to do this was to provide access for small-scale farmers in the Global South to markets in the Global North and to create awareness among consumers that buying fair trade really improves the livelihoods of marginalized farmers abroad.

"Some products people really know about or associate with fair trade include coffee, cocoa, perhaps artisan crafts, but it's really important to know that the vision of fair trade was not just to create a niche market for politically correct consumers. **It was about a systemic transformation of our global economic trading system and trying to create partnerships for trade that are really based on principles of transparency, respect, accountability, and solidarity.**"

Harriet Lamb, Fairtrade International's London-based CEO, oversees the certification of everything from bananas to soccer balls. "Fair trade is, first and foremost, a grassroots social movement for change," she explains. "We put justice back into trade and try to overturn some of the centuries-old injustices that allow tea workers to end up getting just one percent of the final price we pay for the product in the shops, or that allow cocoa farmers to get just seven percent of the price we pay for our chocolate. We seek to overturn these kinds of injustices and achieve change through public mobilizing, through people talking about fair trade and then through certification."

In terms of coffee, the group works directly with farmer cooperatives—at times even organizing small farmers into larger blocks—to establish **GUARANTEED PRICE MINIMUMS** that protect these farmers from fluctuations in the marketplace. Mary Jo Cook, chief impact officer for Fair Trade USA, says, "What that floor price does is let farmers know that even in periods of high volatility, they're guaranteed a minimum price that will help cover the cost of farming. It lets them take loans out and get access to working capital because the lender can now count on some level of pricing. And it ensures that economic benefit goes back to the farmer."

Can a global movement be built with a bunch of small farmers banded together in cooperatives? Harriet Lamb thinks so.

"We only work with smallholders because, in fact, they also grow the overwhelming majority of many crops," she states. "**Globally, seventy percent of our world food is grown by smallholders. If we're going to make the world food system sustainable and fair, we have to keep smallholders front and center in everything we're doing.** We also believe that those smallholders must be organized because if they're alone they will always be vulnerable in this vast global supply chain."

Lamb suggests we're seeing a "kind of quiet revolution," one that proves there is a living alternative based on putting farmers first by guaranteeing them a fair price for their goods. Why? According to Lamb, it's simple. "The public does have those values. They do want their food to be fair."

The genius of fair trade is that it not only establishes a price floor to protect farmers but also provides a fair mechanism to distribute profits. Most companies that

FAIR TRADE VS. DIRECT TRADE

standardization of specific rules and systems. The goal of 1000 faces is to be the storyteller at the campfire, both simplifying and illuminating the story of coffee.

BEN MYERS SAYS BOTH FAIR TRADE AND DIRECT TRADE HAVE PURPOSE AND PLACE IN THE GIGANTIC WORLD OF GLOBAL COFFEE TRADE. The two terms make special

reference to systems which organize how business is conducted in transferring coffee from farmer to roaster. Fair Trade is the more standardized

and mechanized approach, putting the process before relationship and quality. Direct Trade is a more personalized and nuanced approach, favoring relationships over the

codification + mechanized process (1st)
(relationship/quality come second)
sets standards and rules to improve the lives of growers and producers by paying them an above commodity market "fair trade" price

interpersonal approach (others) + quality assurance
(standardization of rules and systems come second)
creates transparency between the two major producers in the coffee chain — roaster and farmer — and provides price incentives for farmers able to produce higher quality coffee

"Farmers who have the ways and means of producing exceptional coffee deserve to be paid better than fair trade prices."

Ben drinks his coffee black

1. Weighing out a batch
2. Roasted
3. Cooled
4. Brewed
5. Flavor Tested
6. Bagged for Delivery

Kochere from Ethiopia

1000 Faces Coffee
Athens, GA
28 June 2011

make profits reward their executives or owners. When fair trade goods sell at higher prices, a portion from these sales is set aside as a **FAIR TRADE** or **COMMUNITY DEVELOPMENT PREMIUM**.

As Colette Cosner describes, "If you as a roaster buy fair trade–certified coffee beans, for every pound, twenty cents goes back to farmers and workers who vote and decide, 'Should we invest this money and improve in the quality of our crops? Should we invest this money in scholarships so our kids can attend high school? Should we invest this money in clean drinking water and better environmental practices?' The key is there's a predetermined economic benefit for farmers, not for NGOs in America or well-intentioned farm owners of large farms, but the farmers themselves."

It's profit for social good.

DIRECT TRADE takes an inverted approach. While it closes the gap between farmer and consumer, it's less about the farmer and more about the company buying directly from that farmer. Companies define their values, then apply them to the products they **ETHICALLY SOURCE**. For example, coffee roasters in Eugene, Oregon, may use beans from growers thousands of miles away, but they may buy their beans based on values that are important to them, like protecting rain forests. So when you buy a cup of this roaster's direct trade coffee, you're also supporting their interest in protecting rain forests.

Kim Elena Ionescu is the coffee buyer and sustainability manager for Counter Culture Coffee in Durham, North Carolina. Over the course of many employee meetings she helped the company boil its values down to four elements: communication, quality, transparency, and price.

"It's not just about paying certain prices and then communicating them back to the farmer—that's only the transparency piece—or providing good quality for our consumers," she says. "It's the whole idea that we're working on this collaboratively with our farms and being friends. Our coffees are a product of a personal relationship."

Andrew Feldman, executive director of As Green As It Gets, a direct trade group that works closely with Guatemala coffee growers, would add stewardship to the list. "**I think both stewardship of the environment as well as stewardship of the people, people who put labor and time into creating that product, is what determines the ethical sourcing of a product.**"

Direct trade is a one-to-one relationship, a **CONNECTED MARKET** version of "Know Your Farmer." It tells consumers about both the company making a product and the ingredients that go inside. Direct trade designation is often printed right on a product label. However, unlike fair trade, direct trade is not an industry-wide certification. Its definition—and specific values—differ from company to company. Counter Culture, for example, hires third-party auditors to verify the authenticity of its own direct trade products to ensure they're everything their program intends. But does direct trade certification cover everything?

When I ask Kim Ionescu why Counter Culture Coffee left "land stewardship" out of its value equation, she offers an interesting perspective. "I wish that one label, one certification, one program could encompass all the things I feel strongly about as a buyer, all the things I see walking around a farm that make it special. But we couldn't include something like land stewardship in our certification program because it would have required that our farms be audited. For our direct trade certification, we really wanted the focus to be on Counter Culture Coffee as an entity as opposed to something that's focused on the farm and the farmer."

"Direct trade means a lot of things to a lot of different people," Andrew Feldman concludes. "In a town you might hear that direct trade is a local transaction: 'I go down the street to the farmer who lives on the road and buy his products directly.' For other people it's ethical. 'I'm dealing directly with the farmer who toiled out there in the fields. He's the one whose coffee I'm buying.' Or direct trade is about quality. It's saying, 'I'm a big coffee company. I'm buying a million pounds of coffee from a big, big grower. I'm buying directly from them, meaning I have control over the quality of the product. I know where it comes from so I can ensure that I'm getting the best quality of coffee to my customers.' I don't see anyone coming up with a consensus around what direct trade represents because direct trade means so many different things to different people."

While still works in progress, direct trade and fair trade help humanize the supply chain and show how we can create a more equitable and connected food system.

ETHICALLY SOURCED GOODS are "produced and purchased in a manner that demonstrates respect for the people who produce them as well as for the environment. Goods should be purchased through transparent relationships that are built on trust and openness, and producers should be compensated at a level that reflects the value of their hard work and that provides a living income for themselves and their families."
—Andrew Feldman, As Green As It Gets

Delan "Rusty" Perry
Papaya Farmer

Rainbow
papaya

Dennis Gonsalves
USDA ARS Scientist

GENETICALLY MODIFIED

(as defined by the Cartogena Protocol on Biosafety)

any living organism that possesses a novel combination of genetic material obtained through the use of modern biotechnology*

THE RINGSPOT VIRUS NEARLY WIPED OUT HAWAII'S PAPAYA INDUSTRY UNTIL DENNIS GONSALVES HAD AN IDEA.

a non-genetic Kapoho. After field tests conducted by the University of Hawaii proved the Rainbow's resistance to PRSV, seeds were distributed for free to local farmers. "Hawaii's papaya industry has returned to production levels that preceded the invasion. His team created a new transgenic red flesh variety called the "SunUp", then developed the F1 transgenic Rainbow by crossing the SunUp with

GMOs

Consumers in Japan, Russia, Australia, Germany, France, Italy, South Korea, Kenya, and Vietnam know more about their food than Americans do. These countries—and many more—require the labeling of all genetically engineered foods.

GMOS are genetically modified organisms. Scientists take DNA from one species and add it to the DNA of another in ways that could never happen in nature or through natural plant breeding. In the United States, the FDA has approved some of these genetically engineered crops for use, which means **you're probably already eating or wearing GMOs and you don't even know it.**

GMOs have been around for decades. The FDA approved the "Flavr Savr," a tomato genetically engineered to have a longer shelf life, for sale in the early 1990s. Ironically, a product engineered to stay on the shelf longer didn't; it was soon replaced by a conventionally grown variety.

Puna is a farming district on the far eastern side of Hawaii's Big Island. The volcanic soil here is deep brown, almost coffee colored, and rich in minerals. One of Hawaii's leading export crops grows on these gently sloping hills. A long gravel road winds over them to Delan Perry's farm. We stand in his packing shed, a rusted tin-roofed shack without walls on three sides. The back is lined with tools, discarded machinery, and packing crates jammed with intensely perfumed papaya. Perry is a thin, deeply tanned farmer who moved here after Oahu's papaya industry crashed twenty years ago. He produces a knife from his belt and cuts off slivers of pink-fleshed SunUp papayas destined for the Japanese market. The yellow ones are called Rainbows. I ask where these grow and Perry points toward a hill rising beyond the packing shed. "It's a hike," he tells me.

We're joined by Dennis Gonsalves, a scientist with the USDA's ARS program. Imagine your stereotypical genetic scientist. Gonsalves is not what you're picturing. He's jovial and energetic, and dressed in impossibly white slacks and a tasteful Aloha shirt. In 1991, the papaya ringspot virus was discovered right here in Puna. Coincidentally, Gonsalves was conducting field trials for a genetically modified papaya at the same time. He'd begun researching the ringspot problem a decade earlier at Cornell University and was perfectly equipped to confront the challenge facing Puna farmers.

"I think what separated our small research group from other groups—even now—was that we had the data, did field trials, then published scientific papers about our breakthrough," he recalls. "Most scientists would be glad to stop there."

Gonsalves and his team didn't.

He guesses that large seed companies typically spend up to $50 million to develop and get approval to commercialize a transgenic seed. Despite the obvious obstacles—including the fact that they had no money and had never done it before—his small team overcame initial reluctance from the U.S. Environmental Protection Agency (EPA), which had initially classified the Rainbow as a pesticide after concluding that **GENETIC ENGINEERING** or **GE** allowed the plant to kill the ringspot virus. Meanwhile the virus kept spreading. Entire farms on Oahu were pulling up thousands of acres of papaya.

"It was desperate," Gonsalves tells me. "These people were out of time."

With the EPA finally on board, they won the support of USDA's Animal and Plant Health Inspection Service and the FDA. These agencies, as well as government officials, made repeated treks to Delan Perry's farm.

Finally, on May 1, 1997, six years after the virus first appeared in Puna, Gonsalves and his team invited local farmers to a downtown Hilo hotel, then gave their Rainbow seeds away.

For free.

I ask Gonsalves if he has regrets, considering how lucrative the GMO industry has become.

He laughs. "The overriding thing, as a plant pathologist, is that when I first heard of the problem in 1978, I thought of it as that: a problem. My job, as a public sector scientist, is to solve problems."

"So you don't miss the money?" I ask.

He claims he doesn't. "Technology is cold. It's in a lab. What we did was to humanize technology. We did something to help people."

I'm not entirely convinced. That comes later, when Gonsalves tells me about his childhood. He was raised on a sugarcane plantation, then left for the mainland. He stopped first at the University of California, Davis, then moved to Cornell, but Hawaii remained his home. As with most islanders, the place defines him. **His ability to create a genetically engineered papaya was less important than using his knowledge to save a key industry where he grew up. The problem and its solution were both personal and local.**

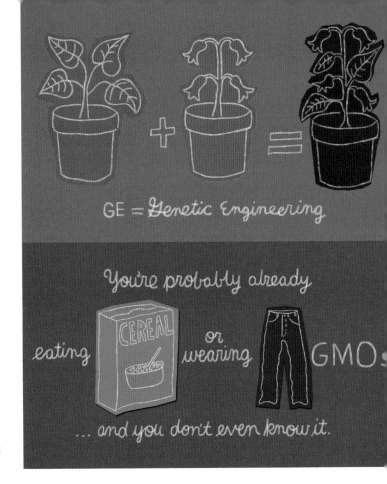

GE = Genetic Engineering

you're probably already eating or wearing GMO... and you don't even know it.

USDA programs supported farmers with seeds until the 1920s (by 1897 they had given away more than one billion seed packets). But when seed companies adopt the same practice as Gonsalves—taking specific plants, isolating select genes in their DNA, then inserting these genes into other plants—they patent their results.

Farmers can't simply purchase GMO seeds. Instead, they sign licensing agreements that heavily restrict their use. Those that fail to follow these contracts—by attempting to save seeds, for example, a practice farmers have followed for thousands of years—often end up in court. Why? Because GMOs are now a multibillion-dollar business. **Seed companies invest millions in seed development and contend that GMO seeds should have the same protection as computer software, vaccines, or any other patentable intellectual property.**

Critics reply on two fronts: first, that for the sake of food security, private companies should not own key aspects of our global food supply; second, they claim GMO seeds are based on publicly available plant genes and the DNA of live organisms. What's next? they wonder. Biotech firms making the same claims to patent human genes and DNA?

Consumers hear that GMOs resist drought, increase crop yields, and require fewer pesticides. These are impressive claims. They're also difficult to verify independently.

John Fagan is the founder of Global ID, an international monitoring group that analyzes GMOs in food and agricultural products. His 2012 study, "GMO Myths and Truths," offers a comprehensive assessment of GMOs from all corners of the scientific community.

"Back in 1995, when I first started talking about it, we could say, 'We know how GMOs are made, there are some risks,'" Fagan recalls. "Now there are hundreds of peer-reviewed scientific studies showing that GMOs are hazardous to our health and damage the environment. You look at that body of knowledge and it's clear. Every GMO, before it's put on the market, needs to be tested rigorously. That hasn't been done in Europe, in the U.S., or in any other country."

That Fagan collected so many peer-reviewed scientific papers critical of GMOs is less surprising than the fact that these papers were produced at all. GMOs are notoriously hard to test.

"That contract farmers sign not only forbids them from saving seeds, it also forbids them from giving those seeds to anybody else, like researchers," Michael Hansen, senior staff scientist at Consumers Union, points out. "That's why most of the leading studies that find adverse health effects tend to have been done outside the United States, often in Europe."

Even though GMO seeds are commercially available in the United States, researchers are required to ask permission from these seed companies before conducting tests. If no permission is granted, there's no legal way to perform a study. Conversely, permission given can be just as easily withdrawn if a seed company becomes disenchanted with the potential outcome of this research.

"It's impossible for third parties to either confirm or disprove the claims these biotech companies make," says Megan Westgate, executive director of the Non-GMO Project, a labeling program that conducts exhaustive audits of a product's ingredients to verify that it's non-GMO. "But there are some studies that have been done, and when you see the initial data coming in, it's certainly understandable why consumers are increasingly alarmed about eating this technology and feeding it to their families."

Should consumers avoid GMO products? Do most industrialized nations, by either banning GMOs or requiring GMO labeling, know something we don't? Should our government step in to sort things out? Proponents of a free-market economy would like market forces to prevail and let consumers decide, but making informed decisions requires transparency. Access to information. Facts. Consumers want to know what they're eating. These concerns don't make them Luddites. They don't demonize technology. Nor are they hippies secretly advocating a new "back to the land" movement. Then what are they afraid of? Maybe they remember DDT. Or read about the lethal outbreaks of mad cow disease, salmonella, and *E. coli*. Maybe they witnessed decades of

legal wrangling—long after the public knew the facts—before the government finally concluded that cigarettes really do kill people.

"The tobacco industry used a systematic approach to generate doubt," John Fagan points out. "The research was there, but they could always find something saying, 'Well, there's some doubts here so we shouldn't have to do anything.' It's the same thing happening with GMOs."

We can't exactly return GMO crops to the laboratory, especially after they've been growing in American fields for nearly twenty years. Isn't the proverbial cat already out of the bag?

"Things get pulled off the market all the time." Michael Hansen contends. He says the time to act is now, instead of waiting for the consequences of GMO use to be known. "We haven't used DDT in twenty-five years but if you take a fat sample from anybody on earth, you'll be able to find DDT or its metabolites. So should the answer be that since persistent organic pollutants are everywhere, we shouldn't take any action against them? You can ban these things and then the situation gets better. It might take a little longer, but they've recalled pesticides and all sorts of products that they first thought were safe."

Another solution might be to simply label foods that are non-GMO and leave the rest to consumers.

Those that feel GMOs pose no risk can continue to buy GMO foods. **Those who have grave concerns, who want GMOs to be evaluated in a laboratory first, not in their intestines, can simply select non-GMO foods.** The marketplace will decide, especially since food producers pay close attention to their consumers.

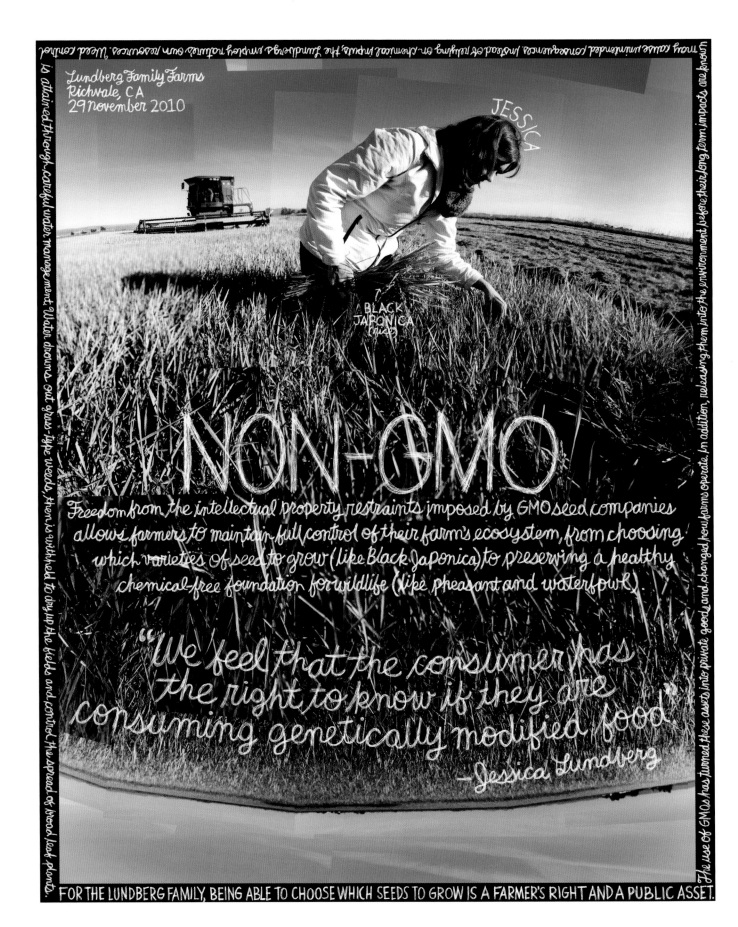

Lundberg Family Farms
Richvale, CA
29 November 2010

JESSICA

BLACK
JAPONICA
(rice)

NON-GMO

Freedom from the intellectual property restraints imposed by GMO seed companies allows farmers to maintain full control of their farm's ecosystem, from choosing which varieties of seed to grow (like Black Japonica) to preserving a healthy chemical-free foundation for wildlife (like pheasant and waterfowl).

"We feel that the consumer has the right to know if they are consuming genetically modified food."
— Jessica Lundberg

FOR THE LUNDBERG FAMILY, BEING ABLE TO CHOOSE WHICH SEEDS TO GROW IS A FARMER'S RIGHT AND A PUBLIC ASSET.

"We're farmers," says Jessica Lundberg of Lundberg Family Farms, whose family has grown rice for three generations in California's Sacramento Valley. "We created a relationship with our customers by selling them products with our name on it, by having conversations with them and asking, 'What do you want in your food? What's important to you? What does quality mean to you?' We've discovered that people want to know where their food comes from. They want to know that their food is healthy and safe. That's something we're very aware of and it's one reason why we support this idea that the consumer has the right to know if they're consuming genetically modified food."

These transparency concerns led Lundberg Family Farms to join the Non-GMO Project.

"We only launched a couple of years ago and we're already at $3.5 billion in annual sales of verified products," Megan Westgate points out. "That's certainly inspiring a lot more companies to get their products verified."

What's the rush? Is it that Whole Foods Market, a national supermarket chain, announced that food producers have until 2018 to disclose the presence (or lack) of genetically engineered food ingredients in their products? Or is it ballot initiatives requiring GMO labeling in states across the country? Either way, meaningful food labels will require meaningful third-party verification. The Non-GMO Project hopes to accomplish this by working with food companies to establish rigorous protocols for GMO avoidance. This includes testing all major GMO-risk ingredients used—from corn to soy to canola—then measuring these ingredients against European Union standards.

The challenge is hardly trivial. Most large food companies have difficulty tracking the source of their own ingredients—genetically engineered or not. They make bulk purchases in hundreds of markets, then consolidate them at centralized receiving stations where they are then reapportioned for distribution to individual production facilities. These supply chains are byzantine and opaque, often making it impossible for these companies to know the precise ingredients in any product, especially when these ingredients may change with each production run.

For these companies, it's not just a matter of adding a few words to their packaging. Non-GMO labeling will require companies to completely revise their supply chain management, then perform audits of all ingredient suppliers to ensure compliance. In short, adding a few words to a product label will transform the entire food industry.

Despite his outspoken support of GMO labeling laws, John Fagan isn't opposed to the use of genetic engineering. He just doesn't think it's used correctly. An example he points to is snorkel rice. Flash floods always present a problem in rice production. The Japanese eat a lot of rice, so a research team led by Motoyuki Ashikari, a professor at the Bioscience and Biotechnology Center at Nagoya University, resolved to address the problem. Genetic engineering allowed them to identify two different flood-resistant rice genes, then marker-assisted selection guided the process of cross-breeding two different rice varieties— a high-producing type and one adapted to deep water. **Basically, genetic engineering provided the DNA road map, showing the scientists what to look for. From there they used traditional seed breeding methods to identify progeny from these crosses that had these desired "high-yield flood-resistant" genes.** The result is rice that, when flooded, will grow up to twenty-five centimeters a day and continue growing until it reaches the water's surface. As John observes, "What you've done is use the rice's subtle, refined regulatory mechanisms and incorporated them using natural methods. As far as I'm concerned that's the way, that's the future for breeding and genetic engineering."

Doug Gurian-Sherman, senior scientist with the Union of Concerned Scientists, agrees. "Many types of crops have been grown by small farmers for centuries and have adapted to a wide variety of environmental conditions and pests," he points out. "We have barely scratched the surface of these possibilities. The risk is that we devote relatively few resources to breeding, perhaps due to our misplaced hope that genetic engineering is what we need. That may come back to haunt us."

LOCAL FIRST, CERTIFICATION SECOND

Francis Lam is an **ETHNOFOODOLOGIST**. He writes about culturally diverse populations that grow food. It's Lam who tells me about Vietnamese shrimpers in Biloxi, Mississippi. I hadn't planned on a trip to the American South, but Lam is convincing. He points me toward Pableaux Johnson, who introduces me to Richard McCarthy at Market Umbrella—a nonprofit that runs Louisiana farmers' markets—who tells me about John T. Edge and Mary Beth Lasseter at the Southern Foodways Alliance in Oxford, Mississippi. Two weeks later I'm standing in a vacant lot where a house once stood until Hurricane Katrina. Now it's a community taro field tended by the folks from Mary Queen of Vietnam Church in North New Orleans (see pp. 220–221). From there I head east to track fishing boats off the coast of Biloxi. I've given myself one month to reach Virginia and it already feels like that won't be enough.

After a few days my assistant observes that we've yet to photograph an organic farmer. I find that to be impossible. Of course these people are certified organic. After all, the most food-savvy Southern minds have shaped our itinerary. I resolve to ask this question to the next farmer I meet.

I receive my answer in Birmingham, Alabama, talking to Edwin Marty at Jones Valley Urban Farm. **I ask if he's certified organic and he says, "Oh. You must be from California."**

Alabama has the fewest number of certified farms in America. Marty claims that while many Southern farmers practice sustainable methods, **ORGANIC CERTIFICATION** in this region faces three challenges. First, many areas lack local distribution systems to connect interested farmers with the consumers who want these organic goods. Second, Alabama farmers don't get enough guidance and support to go organic. He claims local and state agencies promote conventional agricultural solutions that require petrochemical-based pesticides and fertilizers. Third, southern farmers don't want the federal government to come in and tell them what to do with their land.

So, what's the answer?

For Marty, the answer is a principle called **LOCAL FIRST, CERTIFICATION SECOND. He says, "Organic certification or a piece of paper will never ensure you're getting good food. You have to know your farmer."**

When you know your local farmer, you learn about his or her farming principles and practices, but how can you certify that?

Jay Martin, a farmer nearly one thousand miles away in Bivalve, Maryland, thinks he has the answer. He grows dozens of crops and, like Marty, isn't organically certified either. So what certification does he have? **FACE CERTIFICATION**, a direct contact between farmer and consumer that creates an environment of trust and faith.

Trust and faith. Knowing your farmer. These are some of the things to think about when you decide whether to go local or organic.

Or both.

"This farm uses food as a tool for social transformation. It reconnects the community to their food and improves health outcomes in an urban environment."
— Edwin

Jones Valley Urban Farm
Birmingham, AL
26 June 2011

FRANK STITT
FOUNDER/CHEF
HIGHLANDS BAR AND GRILL

"People like me want organically raised food—its better for the land and our environment—but its more important to work with and support local farmers. They put us in step with the seasons and create a valuable bond between farmer, chef, and community."

EDWIN MARTY
FOUNDER/FARMER
JONES VALLEY URBAN FARM

Edwin came to town with the vision of reconnecting this urban community with sustainably grown food so he reclaimed a small plot of land and started growing herbs lettuces and vegetables for his community.

LOCAL FIRST,
CERTIFICATION SECOND

Organic certification or a piece of paper will never ensure you are getting good food. You have to know your farmer.

WHY DOES ALABAMA HAVE FEWER CERTIFIED ORGANIC FARMS THAN OTHER PLACES IN THE COUNTRY? THREE OBVIOUS REASONS: 1) It's hard to grow food in Alabama without synthetic chemicals, because of the high humidity; 2) insect predation; 3) poor soils. THREE LESS OBVIOUS REASONS: 1) Black farmers were intentionally excluded from USDA funding for decades, so why would they now trust the USDA to make their work more valuable? 2) The good ole boy network of farming in Alabama continues to view organics as a "hippy" thing. 3) County extension agents only offer research-based advice and the land grant universities have not been incentivized to do such research.

THE CROP RENT CALCULATOR

An agile equation used to determine a crop's true value within a greenhouse's diversified year-round growing system

Stone Barns Center
for Food & Agriculture
Pocantico Hills, NY
26 June 2012

COST OF PRODUCING 1 LB OF MOKUM CARROTS

L = 50ft = length of each plant bed
T = 50 = # of days crop grows from seeding to harvest
0.33 = value of 1 square ft. of growing space in greenhouse (calculated by factoring the total annual expenses of operation including labor, capital value of structure, materials, seeds, tools, fuel, and electricity then dividing this number by total productive days in the year)
2.5 = width of the bed (2.5 ft.)
75 = average weight per bed for mokum carrots grown here

the mokum carrot

$$\frac{L \times \left(\begin{array}{c}\text{V of Bed ft}\\ 0.033 \times 2.5 \\ = 0.0825\end{array}\right) \times T}{75} = \$2.75 \text{ per pound}$$

(50)   (50)   (and it's sold for \$4)

HOW DOES JACK ALGIERE OF STONE BARNS DECIDE WHAT TO GROW IN HIS GREENHOUSE? A CROP RENT CALCULATOR.

How extensive crop selection works together to form a viable, ecologically sound growing operation. 3) Determine crops that work best within the ecological fabric necessary to maintain long-term soil health.

It allows him to: 1) Track seasonal growing trends; 2) Adjust pricing to improve economic efficiency and better understand

PART III
UNCONVENTIONAL AGRICULTURE

"I worked all the way through high school doing different things. I remember working in these greenhouses filled with geraniums, all pretty much the same variety. This black fungus started to grow around the base of these plants. I turned to my boss, he was really my mentor, an incredible man and a brilliant grower, and I just remember saying, 'I see this thing on the plants, what do we do?' And he said, 'I got this stuff. We'll use this fungicide and take care of it, but I need you to put on that suit and duct-tape your wrists. Duct-tape your ankles and duct-tape that mask right to the hood of your suit. Then we'll go in there and we'll spray these things. Just do it all.' And I said, 'Why do we need to do that? Can't we just remove these couple of plants?' And he said, 'Oh, no. That's not how it works. If it's on one of them, it's on all of them. It's like wildfire. There's no stopping it. This is the only way to deal with it.' I just couldn't believe it. I couldn't believe what we were about to do. And I said, 'Why do we choose to grow these things if we know this is a potential risk?' Usually he gave me answers so in depth about how plants grow and what they need and all these things, but the answer he came back with that day was, 'Well, this is what the customer wants.' You know, someone so brilliant was basically under the thumb of an entirely uneducated community. They wanted things and he didn't feel it was necessary to tell them what kind of health risks he was exposing himself to, what kind of environmental risks he was taking with the system, just the whole thing was not good. It was like a spear in my heart."

—Jack Algiere, Stone Barns Center for Food & Agriculture, Pocantico Hills, New York

SEEDS

One moonless February night I walk through a freshly cut hay field in Sebastopol, California. My destination is a single incandescent bulb. It illuminates the front door of the Sebastopol Grange Hall. Inside this modest building people have gathered, their heads bowed in deep concentration over rows of folding tables. On each are clusters of carefully marked glass jars, which they inspect closely, even hold up to the light. After some consideration they may gently open their lids, take a sniff, then dip some of these precious contents into small envelopes that are thrust into their now-bulging pockets, but not before scribbling cryptic admonitions like "careful—a real climber" or "full sun only" or "must have drainage."

This is a **SEED SWAP**, and the people here have come to share seeds—and stories—before spring planting.

"The world is crying out for more community," says Sara McCamant, cofounder of the West County Community Seed Exchange and organizer of tonight's event. **"People are coming together and sharing resources, learning what grows well from one another. We're strengthening their ability to grow their own food, because when you share seeds you also share knowledge."**

In earlier times, seed swaps were a yearly occurrence in any farming community. Farmers grew plants, saved seeds from those that performed best that season—or tasted sweeter or produced more beautiful flowers—then replanted them the following year as the cycle of life repeated. These seeds were **OPEN-POLLINATED**, meaning they were "true-to-type" and would produce plants just like their "parents" the following year. Over time each successive generation adapted to its geography's unique climate, temperature, soil, pests, and even plant viruses. **Open-pollinated seeds were a cultural record. They stored vital information about a community, passing it from one generation to the next. Civilizations rose or fell—and even went to war — depending on the success of these seeds.**

West County Community Seed Exchange
Sebastopol, CA
26 February 2013

resources in a world that is crying out for more community; they're strengthening people's ability to grow their own food.

...of seeds, people are coming together and sharing

SEEDS GROWN AT COMMUNITY SEED GARDEN

1. Tarbais
2. Rio Zape
3. Golden Buddha
4. Scarlet Runner Bean
5. Aunt Jean's Pole Bean
6. Three Grex Beet

7. Curly Red Mustard
8. St. Valery Carrot
9. Red Kuri Squash
10. Thelma Sander's Sweet Potato Acorn Squash
11. Brightest Brilliant Quin
12. Floriani Polenta Corn

SARA MC CAMANT SAYS THE ONLY REASON WE

When you share seeds,
you share knowledge.

again and given them to someone else. Seeds are really the first link in a local food system. Be-

SEED SWAP

a community of farmers and gardeners gathering to exchange seeds

SARA

They have passed them around, shared them, improved them, grown them, improved them

THE AMOUNT OF DIVERSITY AVAILABLE TO US RIGHT NOW IS BECAUSE PEOPLE HAVE SAVED SEEDS.

PRESERVATIONISTS

Seeking out, collecting and propagating heritage varieties to secure their enjoyment and appreciation by future generations

VIRGINIA GREENING
firm + crisp with full-bodied flavor + hints of vanilla

ANTIQUE APPLE
One with a documented history of 100 years or more

HOMESTEAD APPLE
An apple passed down from one generation to another (also called "heirloom")

Big Horse Creek Farm
Lansing, NC
2 July 2011

Sought them, brought and preserve apple trees. With this knowledge they moved to the mountains of Ashe County, where they now raise almost 400 different varieties. If they are "collectors and propagators" on a mission to seek out, collect and preserve the precious few heritage apple varieties still in existence

Having always believed that heirloom fruits and vegetables are critical in an agriculturally homogenous society, they become protégés of Lee Calhoun. He

APPLES WERE PIVOTAL TO THE ESTABLISHMENT OF AMERICAN AGRICULTURE BY FRONTIER SETTLERS (BUT WITHOUT RON AND SUZANNE JOYNER THEY'D NOW BE MOSTLY FORGOTTEN)

"America's founding fathers were 'founding farmers' who recognized that we lacked the appropriate seeds to feed a growing nation," explains Matthew Dillon of Seed Matters, a nonprofit dedicated to organic seed preservation. He tells me stories about how Thomas Jefferson smuggled rice seeds out of Italy by sewing them into his coat lining, and how our earliest ambassadors were ordered to collect seeds from each port of call. "Seeds have been a public natural resource that has been shared among people for over ten thousand years," he continues. "In our republic's early days, the U.S. Patent and Trademark Office even grew out seeds, documented their collections, and sent out letters to agricultural journals announcing their discoveries." More important, they gave seeds away—more than a billion before the close of the nineteenth century. To hear Dillon tell it, spreading seeds was a patriotic act.

I'm reminded of Big Horse Creek Farm, the remote, off-the-grid hilltop farm Ron and Suzanne Joyner settled thirty years ago in Ashe County, North Carolina. They're **PRESERVATIONISTS**. Or as Ron explains, "Our mission is to seek out, collect, and preserve these last remaining heritage apple varieties so future generations can enjoy and appreciate a part of our history."

And he's not kidding. The hills below their cabin serve as a quiet refuge for nearly four hundred **HEIRLOOM** apple trees grown from scionwood (cuttings) gathered on various journeys across the United States. If not for them, many rare apples, including the Virginia Greening, a thick green-and-yellow-skinned apple from the 1700s, might no longer exist.

Ron doesn't have one favorite apple—that would be akin to asking a parent to choose a favorite child. Instead, he has many. As he explains, "We have our favorite fresh-eating apples—the Myer's Royal Limbertwig, the Husk Sweet, and the Tompkins County King. We have our pie apples—the Porter and the Kinnaird's Choice. Then come our apples for applesauce and apple butter—the Yellow Transparent and the Wolf River. We also have cider apples—the Grimes Golden and the Golden Russet. Oh, and the long-storage apples—the Yates and the York Imperial."

Later that afternoon we climb the hill back to their cabin and share a glass of water while I leaf through a massive tome called *The Apples of New York.* Printed in 1905, it features hundreds of dreamy, semi-erotic illustrations of apples with names like "Westfield Seek-No-Further" or "Esopus Spitzenburg," a Jefferson favorite.

Don't bother looking for any of these varieties at your local supermarket. **We've gone from a nation that prized AGROBIODIVERSITY—that treated seeds as something vital to our national interest—to a country that resorts to terms like HEIRLOOM and HERITAGE to underline the rarity of**

OPEN-POLLINATED: When pollination occurs naturally. Because there are no restrictions on the flow of pollen between individual plants, open-pollinated plants are more genetically diverse.

"An **HEIRLOOM** seed has been passed along for generations within a family or community. What attracts me to growing heirloom varieties is not just that they are heirlooms. I like them because they are uncommon in my area, and beloved and prized in their own home country.

I love to find a distinct herb or a cucumber or any type of vegetable seed, grow it, and find someone who remembers eating this distinct variety when they were children."
—Annabelle Lenderink, Star Route Farm

foods that were commonplace a few generations ago. But even these terms are too imprecise for Glenn Roberts, founder of Anson Mills, in Columbia, South Carolina.

"The term *heritage* can be really good or it might not mean anything," Roberts surmises. "It's better for marketing and it's better as a general cultural concept than it is as a botanical reality."

He takes similar issue with the use of *heirloom*, which generally refers to seeds fifty years or older. He thinks we need to go further back, starting before the advent of industrial agriculture. "The public's perception is 'this is what great-grandma used to do,'" he continues. "She kept her seeds in the jar and planted them out, but that's not necessarily what *heirloom* means anymore. That half-century mark now brings us back to the Green Revolution, so if you're talking about heirloom grain in the 1950s, you're already looking at dwarf crops. You're looking at crops that are highly suspect today because of their densities, their mineral depletion, the fact that they are chemically farmed, which impacts their genetics."

The story of how our seeds lost their innocence is both interesting and obvious. We evolved from an agricultural system based on biodiversity to an industrialized model that grows much less diverse crops. The antecedents of this shift date back to the 1920s, when trade associations exerted enough leverage to radically curtail the USDA's public seed program. The responsibility to breed and distribute seeds shifted to the private sector and land-grant university plant breeding programs, whose research primarily supports farmers and food processors.

"A **HYBRID** seed is artificially produced, usually from a controlled pollination with pollen from two very diverse parents. Seeds from subsequent generations are all over the map and cannot be saved, which breaks the cycle between people saving their own seed and adapting a variety to their own climate and location."
—Annabelle Lenderink, Star Route Farm

During this period plant breeders popularized the use of **HYBRID** seeds by combining qualities from different plant varieties. These new hybrids produced perfectly uniform crops but their "false-to-type" seeds could not be saved. Farming quickly evolved from a diverse system to one of rotating monocultures, and since hybrid seeds could not be saved, the art of saving seeds at harvest and storing them until next spring's planting nearly disappeared.

In 1980, passage of the University and Small Business Patent Procedures Act—often referred to as the Bayh-Dole Act for the senators who sponsored it—presented universities with a new challenge; instead of relying on federal grants, these institutions were now asked to find outside funding sources and to secure patents and royalties as additional sources for revenue.

As Matthew Dillon explains, "In order for a program at Iowa State University or Purdue University to receive funding to educate a new generation of public plant breeders, they have to do basic R&D for the private sector as opposed to serving the public good and the needs of regional agriculture."

Large seed companies are interested in technology advancements that boost their bottom line, which means their **GENE JOCKEYS** are more likely to focus on patentable breakthroughs related to the largest agricultural markets—corn and soy, for example—rather than breeding the perfect carrot for Georgia's humid clay soils (or tomatoes that will grow in the cool summer climate of Petaluma, California). And by focusing research on large crops, they can exert control over their commercial production. As Micaela Colley of the Organic Seed Alliance points out, "When you invest all your effort into plant breeding being done by a few people, what you're really investing in is breeding for the needs of a centralized food production model. We're not advocating that every farmer needs to breed his or her own crops, but we believe in creating regional networks of farmers, seed companies, and university-based support, plus independent and public plant breeders that all work together to address the needs within a region."

I find that near Boulder, Colorado.

HEIRLOOM VS. HYBRID
heirloom: open pollinated seeds saved and passed from one generation to the next
hybrid: artificial, controlled pollination developed for large scale agriculture
1) rossa di milano; 2) tropeana lunga;
3) tropeana tonda; 4) torpedo
20 July 2009
La Tercera Farm
Bolinas, CA
Annabelle
4.
2.
3.
1.
me
ANNABELLE GROWS HEIRLOOM ONIONS

RICHARD SHANAN

Abbondanza Organic
Seeds and Produce
Boulder, Colorado
12 August 2010

beets
seeds
greens

COLLECTING BEET SEEDS
IS A BIENNIAL PROCESS

This is the fourth generation of
Shiraz beets grown on this farm.
They grow out rootstocks, harvest
them in Fall, store them, and then
replant in Spring. The replanted
stocks bloom. As these flowers
mature, their seed heads are
gathered.

SHIRAZ SEEDS ON THIS FARM
ARE SELECTED BASED ON THE
FOLLOWING CRITERIA:

1. color
2. maturation
3. size
4. strength of greens
5. flavor
6. endurance

THE BIO-REGIONALISTS

"We recognize that our bio-region, our domestic eco-culture has a unique set of circumstances defined by climate and soil.
Some things flourish here while others are diminished in life force. We seek out seeds that thrive. We love them,
live from them, share in their thriving. We also strive to find things marginal and adapt + strengthen them
in order that there be more biodiversity for us to live from (which runs in opposition to the principles of monoculture.")

RICHARD AND SHANAN ARE SEED COLLECTORS AND BIOREGIONALISTS: they believe that seeds adapt, that plants and animals make decisions based on a region's living biology. They grow and gather in soil with specific minerals and bacteria, defined climate, and pests which impact a seed's regenerative capabilities. By virtue of this address, their seeds are BIO-REGIONALLY adapted to their local environment... it's a kind of deep locavorism.

Shanan Olson likes chickens, which is a good thing because Abbondanza Farm has hundreds of them. They're everywhere. Aside from chickens, Olson and Richard Pecoraro grow enough produce to run a small CSA for residents in nearby Boulder and Longmont. We explore a few greenhouses, then walk the perimeter of their farm, passing acres of well-maintained row crops. One area between the rows is conspicuously overrun with weeds and large brush. It's like they

simply gave up here, maybe embraced the fact that sometimes you can't farm every square inch. It turns out that Pecoraro is most proud of this forsaken little patch. Instead of harvesting, he's let these plants go to seed, or what Olson calls "go full circle." We wade into a jungle of peas, Swiss chard, and amaranth as Pecoraro picks a beet seed head, crumples it in his palm, and shows me the seeds inside.

"These are interesting because beets require a biennial process," Olson tells me. "We grow out the rootstock, harvest them in the fall, store them, then replant in the spring. When these flower we can finally gather the seed heads."

A few minutes later we stand in their warehouse. The raw timber shelving along each wall is jammed with seed sacks bearing cryptic hand-scrawled names like Mexican Red, Dragon, and Japanese Imperial Long. Pecoraro has collected these for years but now focuses on bio-regional seeds, open-pollinated varieties adapted to the low rainfall and heavy clay soils that define the area's unique Front Range geography.

Vandana Shiva is outspoken in her defense of **BIODIVERSITY** but she won't let me turn it into a picture. For weeks I call across twelve time zones to explain my idea: I want to create a photograph of her sitting on the edge of bed in a nondescript hotel room, suitcases at her feet. They burst open and their contents spill across the carpet and over her shoes. Rice seeds. Thousands of precious rice seeds that have been painstakingly saved by Indian farmers. I want to call her picture "The Seed Ambassador," because she spends most of her life away from New Delhi speaking at conferences, hearing the pleas of peasant farmers, educating educators, admonishing the CEOs of large corporations, and inspiring heads of state. Vandana has captured—more than any other individual on the planet—the urgency of preserving our global seed supply. I tell her all this can be contained in a single photograph but she remains unconvinced.

Her reticence becomes clear after we meet. The only photograph that makes sense is one taken in her natural element, but since New Delhi is seven thousand miles away, a group of earnest **SEED SHARERS** in Eugene, Oregon, will have to suffice.

The Lexicon of Sustainability, the project I codirect, is about finding solutions, not problems. In such a conversation most agribusinesses have no place. They are, like Lord Voldemort in the Harry Potter books, "The Companies That Must Not Be Named." Still, it's impossible to speak with Vandana without talking about **TCTMNBN** because theirs is a battle fought across many fronts, both in the media and on the ground, which means any conversation with Vandana has to start with . . . zucchini.

"I know you often speak about the importance of biodiversity, but do we really need ten different types of zucchini at the grocery store?" I ask. "When I was a kid—I had an Italian grandmother—she had ten different types of zucchini growing in her garden, and—"

"You need ten types of zucchini."

"Really? Why?"

"You need ten types of zucchini because the zucchini has a variety of different relationships with the soil, the insects, and the pollinators," she patiently explains. "It also has a whole different relationship with you in terms of your diet. Every variety has a different kind of conversation with different cells of the body. Now that's the subtlety of diversity, of the level of the complexity of our body. The complexity of the ecosystem is something gene-shooting ignorant warmongers cannot understand."

"You mean companies that patent seeds?" I ask.

"Seeds are created by seeds, not by [TCTMNBN]," she replies. "All that they manage to do is shoot a toxic gene like the Bt gene or a Roundup gene through a gene gun. Normally that action of introducing a toxic gene would be considered as pollution of the seed, not its creation. **They put in a toxic gene and say, 'I created this life. I'm the creator of corn. I'm the creator of soya. I'm the creator of banana. I'm the creator of canola and everything I touch.' I call it the biggest 'creation myth' of capitalist patriarchy,** because capitalist patriarchy assumes that when capital touches life—including when it touches to extinguish it—that's when creation begins. This is necessarily blind to the creativity, intelligence, and work of nature; the creativity, intelligence, and work of men; and the creativity, intelligence, and work of the peasants."

"I know seed security is very important for your organization . . ."

"**SEED SECURITY** is important because everything begins with seeds. The food system begins with seeds. If seed is in the farmers' hands, agriculture is viable. If seed is in [TCTMNBN's] hands, agriculture becomes unviable. I saw this happen with their entry in India and their takeover of the cotton sector."

"Can you explain how that happened?" I ask.

University of Oregon
Republic of Eugene
3 February 2012

"Either an alternative
food system will be created
or the Americans will kill
themselves, either from
not having food or having
bad food."

— Vandana Shiva

Vandana

SEED SOVEREIGNTY

The farmer's rights to breed and exchange diverse open source seeds which can be saved
and which are not patented, genetically modified, owned or controlled by emerging seed giants

It reclaims seeds and biodiversity as commons and public good.

TO VANDANA SHIVA, THE NOTION THAT CORPORATIONS CAN OWN THE PATENTS RIGHTS TO SEEDS IS NOTHING BUT A CREATION MYTH

Libraries offer, preserve and lend seeds. Seed lending libraries do the same by their seed stewardship and sharing heritage seeds, and by showing a seed library at... a real library... they reach a fuller spectrum of the community. Our access to local seeds and our knowledge of seed saving has been largely lost this past century. Public

Seed Lending Library

"Saving seeds preserves our biodiversity, builds our local resiliency and maintains our food sovereignty."

WHAT KELLI GREW
lemon cucumbers, zucchini, Swiss chard, dinosaur kale, tomatoes, sunflowers, sunset runner beans

KELLI
Urban Farmer

Richmond Public Library
Richmond, CA
30 August 2013

TERY
Book Librarian

DEVYN
Library Card Holder

SEED LIBRARY

A place where seeds and information are freely shared for the benefit of everyone in a community

REBECCA
Seed Librarian

FREE PUBLIC LIBRARIES REVOLUTIONIZED ACCESS TO BOOKS AND KNOWLEDGE. CAN SEED LIBRARIES CHANGE WHAT WE EAT?

"They do three things to make sure the entire market is theirs. First, they destroy the farmers' SEED SOVEREIGNTY. Eighty percent of seed in the 1990s was farmer-owned. How did they destroy this seed? By telling them that their 'primitive' seed was unproductive and they would bring a 'miracle' seed. They did what they called a 'seed replacement' and brought in their GM seeds. Every farmer thought, 'I'm giving my seeds up but if I don't my neighbor will have it or the next village will have it.' This happened in thousands of villages at the same time until their seed ran out.

"The second thing they did was lock in local companies. Sixteen Indian companies are locked in with [TCTMNBN] and they can't sell anything but their products.

"The third is mysterious. The public breeding suddenly stops. It just stops. **Only three kinds of seeds have historically been available: the farmer's own seed supply, the public seed from public breeding, or private seeds. All three options get locked in by [TCTMNBN], so only their seed is left.** We used to have fifteen hundred varieties of cotton adapted to different climates. Now, if there's a drought, there's a failure. If there's untimely rain, there's a failure. Indigenous seeds were adapted to all of those variables.

"Their cotton fails more often so farmers have to buy more often but they don't have that kind of cash. They're encouraged to take seeds on credit. The seed company agents say, 'Just put your thumbprint on your land papers and here's some seeds for credit.' Now you have farmers indebted for seeds, indebted for pesticides, with crops failing more often. Within two or three years—instead of becoming millionaires as they were told—they lose their lands. That's the day the farmer will go to his field quietly—because that land is his mother's—and without telling anyone, he will drink a bottle of pesticide and end his life. That is the tragedy that these monopolies on seed have created. It is genocide."

Each successive generation of seeds learns from its environment; these experiences are stored in a genetic repository and called upon as these seeds continue to adapt. When they're gone, that genetic record of adaptability disappears with them, which heightens the relevance of Vandana's cautionary tale. As Brian Campbell, associate professor of anthropology and environmental studies at Berry College, in Mount Berry, Georgia, observes, "Everyone is entitled to have access to seeds but we are seeing a trend in which governments and multinational corporations are attempting to prevent widespread access to these seeds because food is power and these seeds are the fundamental unit of that power."

Global concern about the loss of SEED BIODIVERSITY has led to the creation of SEED BANKS. Locking these seeds away is one solution. Campbell has another: He plants them. "I don't think a seed bank for conservation's sake is as effective as a more in situ method where you create new agricultural traditions." To achieve this goal Campbell gives his seeds away, along with stories about what these seeds once meant to growers. "The story might be quaint," he admits, "but it might also encourage them to keep growing it and pass it along to other people and continue that story. It's all about independence."

BIODIVERSITY VS. MONOCULTURE

I arrive at Knoll Farms in Brentwood, California, before dawn because Rick Knoll says his farm looks best at that hour. At this hour the chickens are still asleep in the trees, and while some roosters do crow, it's halfhearted. The ten-acre spread, about forty miles east of Berkeley, more closely resembles an oversize backyard garden than a regular working farm. Rick and I exchange quick hellos, our voices scarcely above a whisper, then start out. Even in near darkness, my problem is immediately clear. No camera lens is wide enough to properly capture the densely packed, exuberant profusion of trees and plants now enveloping us. We walk between rows of Brown Turkeys—massive figs that hold up well in the oven yet lack the characteristic sweetness of their more revered cousin, the Black Mission. We pass through a cardoon jungle, finding agreement on my recipe for that highly undervalued vegetable—boil, peel, bread, then fry its celery-like stems. We duck our heads under rows of apricot trees, skirt a cluster of Kadota figs, then carefully navigate between clumps of lavender. This is what **BIODIVERSITY** looks like. It's chaotic. Messy. A cacophony of smells and color.

After visiting hundreds of farms I've developed a simple rule: When I see cows or goats, I pay close attention. If they approach the farmer, it says all I need to know. If the farmer grows produce, I look for birds. Birds mean small little creatures. Small little creatures mean insects. Insects mean the farmer probably doesn't abuse his pesticides.

As the sky around us lightens I notice birds circle overhead then settle in the nearby trees. They watch as workers appear. Rick steps inside the packing shed, goes over deliveries with his driver—they service a particularly loyal group of restaurants and farmers' markets in a fifty-mile radius—then maps out the day's pick and pack. A few men carrying ladders and pruning shears disappear into a grove of nearby fig trees. Some women gather baskets and head off. I follow after them.

I'd like to make a photograph about organic farming, maybe explain Rick's **BIODYNAMIC** practices, but the scenery puts me at a loss; its hidden meaning eludes me. I hand him a few sprigs of rosemary, then ask him to pose. Admittedly, it's a meaningless, decorative photograph, but I've come all this way. I have to leave with something. I frame up and take a step back. Another step. Another. Then the ground behind me softens, throwing me off balance as my boots descend into velvety soft earth. I turn around and find myself staring into a vast, open expanse of nothingness, at perfectly straight rows carved into the soil with machinelike precision—because machines made them. These extend, emphatically, for miles.

MONOCULTURES are agricultural systems that focus on growing single crops; they rely on uniformity and are often practiced at a massive scale. Agriculture in Iowa exemplifies this approach. Corn and soy extend in seemingly every direction, with their infinite rows only impeded every few miles by one-lane country roads veering off toward the horizon. Large agribusinesses develop chemical-based fertilizers to increase the yields of these crops. They also make chemical-based herbicides to kill weeds and chemical-based pesticides to take care of bugs. Industrial farming requires an "in for a penny, in for a pound" approach.

BIOTECHNOLOGY plays an increasingly pivotal—and contentious—role in industrial agriculture, too. In fact, the companies making seed are often the *same* companies making fertilizer, pesticides, and herbicides. One of them (remember "The Companies That Must Not Be Named" or TCTMNBN?) sells genetically modified corn seed; you use their herbicide in concert with their seed, the notion being that the herbicide will kill all the weeds while leaving the GM corn alone. This greatly simplifies farm decisions for most commodity crop growers—it's one-stop shopping—but reliance on these chemical-based solutions doesn't come without problems; herbicide-resistant super-weeds, for example, now appear in agricultural fields across the country. They will eventually require the next generation of even more powerful herbicides. This elaborate cocktail of petrochemicals will beget even more adaptive super-weeds as the vicious spiral continues.

BIODIVERSITY VS MONOCULTURE

weeds + insect crops + critters + soil = an integrated pest well allowing plants to set regulate— (no pesticides needed)

Rick grows only organic fruits and vegetables

← rosemary (harvested year round)

Knoll Farms Home of Tairwa Knoll Brentwood, CA 3 December 2010

AMAYA

nothing grows here (unless Rick's neighbor says so).

growing a single crop over a vast amount of land increases the risk of pests, disease and specialized predators, which conventional farming combats w/ pesticides, herbicides and fungicides.

THE CONVENTIONAL FARMERS NEXT DOOR "CALL RICK'S ORGANIC METHODS "DIRTY FARMING" (THEY'RE "CLEAN"). Each winter their fields sit idle for months at a time. Since no cover crop is planted (this process returns nutrients to the soil and increases soil fertility), the soil remains exposed to the elements. Wind erosion will carry some of this precious top soil away, and in so doing releases carbon back into the atmosphere.

SOIL AMENDMENTS

Farming removes minerals and nutrients from the soil. Amending soil balances its chemistry and improves its physical properties (water retention, drainage, permeability, water infiltration, aeration + structure) so plants and animals can derive the greatest nutritional benefits.

Greenbranch Farms
Salisbury, MD
8 July 2010

TED

Soil amendments used at Greenbranch:
Agricultural lime, Compost, Gypsum, Boron, Rock powder (granite, humalite, kelp, sul-po-mag, sulfate of potash)

Aragonite (made from ground, compressed deposits of sea shells and corals)

Some seeds are naturally unbalanced. Some soils lack minerals such as boron, iron, or zinc (soils). Some soils lack fertility. Where human intervention plays its role in order to fully optimize the potential of the grains to grow. They become dependent on what we supplement it with.

WHEN TED TOOK BACK HIS FAMILY'S FARM AFTER IT HAD BEEN SUBJECTED TO 25 YEARS OF INDUSTRIAL CORN AND SOYBEAN PRODUCTION, HE DISCOVERED THAT THE SOIL WAS INERT.

The industrial footprint took him dearly to make the necessary improvements. One of the main challenges to operating a successful organic farm is to fix soils that are heavily depleted by conventional farmers who have mined their top soil of vital nutrients. He says it is nothing short of amazing to watch the soil come back to life and see crops thrive in healthy balanced soils. Soil fertility is the foundation of any successful farm.

SOIL FERTILITY

Ted Wycall's grandfather was a Maryland farmer but his dad was a dentist. With no one left to work the land, a series of tenant farmers took over. For the next twenty-five years they planted a cycle of conventionally grown corn and soy. When Wycall was old enough, he asked his grandfather for the land back. He did soil tests, eager to see what he'd be working with.

A sustainable agricultural system is one that renews itself. If it's extractive, if it takes out more nutrients from the soil than it puts back, if it depends on finite resources, if it has external costs—with real human and environmental consequences—then it will not survive. This is simple mathematics. Or as Canada's Commission on Conservation put it back in 1915, "Each generation is entitled to the interest on the natural capital, but the principal should be handed on unimpaired."

For Ted Wycall, the "principal"—in this case his soil—was not "handed on unimpaired." In fact, tests showed that twenty-five years of conventional farming had left his soil lifeless and practically inert.

Wycall's story represents the fundamental challenge now facing ambitious farmers across the United States: how to transition land from **CONVENTIONAL FARMING** back to its **ORGANIC** or **"PRE-CHEMICAL"** state. It took eight years of patient research on Wycall's part—interspersed with dark periods that left him paralyzed by self-doubt—before his soil came back to life. The secret was a witches' brew of **SOIL AMENDMENTS** that included aragonite, lime, compost, gypsum, boron, rock powder (granite), humates, kelp, and sulfate of potash. To that regimen he also added compost, cover crops, and straw mulch. Farm animals also contributed to Wycall's fertilizing program.

We spend the afternoon among pastured poultry, grass-fed cattle, and pigs that roam freely—trotting after us like dutiful children—as we explore the forests that line his farm on three sides. Wycall now sells his produce and meat from a farmstand by the front gate and operates a CSA for local families. I wonder whether Wycall's customers know the whole story, what it took to turn this farm around, and what that one reclaimed farm means for the rebuilding of a diverse local food system.

"Farming is one of those things that has recently gained the attention and interest of a large part of the population; everyone thinks, after reading a few books or seeing a few documentaries, that they understand what it is like to farm. I will be the first to tell you that no one will ever understand what this business can inflict on a naïve and well-meaning human being until he has actually spent more than a few years doing this work. It is work that will test your strength in every single aspect of one's humanity. People need to realize that Mother Nature will punish you severely and unfairly many, many times, even though you have done nothing wrong and are only trying to make things better for people and nature. She will kick you when you are down, spit in your face, and the only thing you can do is pray that she doesn't kill you.

"On the positive side, there are many rewards to be had from this work, though they will probably not be financial in nature. I am sticking with farming, but I will not be one to paint a picture of farming that is rosier than reality. All farmers will get burned badly from time to time and they had better be prepared for that. I was not ready to learn that lesson when I began, but I have learned it now and it gives me great humility and enormous respect for all my colleagues, organic or conventional, that take the lickin' and keep on tickin'."
—Ted Wycall, farmer

"Each generation is entitled to the interest on the natural capital, but the principal should be handed on unimpaired."
Canada's Commission on Conservation (in 1915)

(early September), Cresthaven + J.H. Hale (mid-September), and O'Henry (late September))

...t), Rhohaven + Rosa + Coronation (late August), Flouringtan (late August), Alton + Bluewington...

BUILDING
SOIL FERTIL...

Steve Ela is an organic farmer who...
crops to enhance soil fertility on his farm...
deep-rooted perennial legumes like alfalfa a...
white) and vetch which provide biomass earl...
(when the peach trees need it most). They also...
1. suppress weeds; 2. build organic matter i...
soil by bringing nutrients up from deeper soil...
and loss of top soil; 5. perennials don't require...
(continued cultivation disturbs tree roots); and cov...
soil bacteria and microorganisms that support earth...

PEACHES REQUIRE
1. hot days
2. cooler nights
3. lots of fresh
   mountain snowpack
   water (the Rockies)
4. good soil

STEVE ELA'S GREAT GRANDFATHER FIRST PL...

Scientists have long understood that plants grow best when soil has the proper complement of sixteen nutrients. Of these, nitrogen, phosphorus, and potassium—a divine trinity collectively known by its atomic symbols, NPK—are the most important.

For plants to absorb nitrogen (N), farmers have always added composts, crop waste, and animal manure to the soil. These increased **SOIL FERTILITY** and resulted in better harvests. Earlier cultures, especially in Japan, even used human feces as fertilizer. A respectful dinner guest—after ingesting his host's valuable nutrients—would often deposit **NIGHT SOIL** in grateful recompense before departing.

But farmers weren't nearly as efficient as a German scientist named Fritz Haber. In the early 1900s his Haber process resulted in the production of ammonium nitrate, a chemical that fixes nitrogen in the soil. Its primary ingredient? Natural gas. Ammonium nitrate dramatically improved crop yields, and as people learned in World War I, also made very effective weapons.

When the war ended, chemical companies that had assembled laboratories and ramped up production to manufacture large quantities of ammonium nitrate–based explosives now turned their attention back to agriculture. **Farmers who were once guided by the lunar calendar, who carried their knowledge from one season—and one generation—to the next, now received instructions printed on the side of a fertilizer bag.**

Today, half the world's food production depends on a fertilizer whose main ingredient is natural gas. These chemical fertilizers and herbicides are the foundation of conventional agriculture, and a major factor in the public's perception that relative food costs have lowered over the last one hundred years. But the problem is that the overuse of chemically produced nitrogen pollutes our soils, our drinking water, our waterways, and our oceans. These **EXTERNAL COSTS** make the real price of our industrialized foods higher—and more dangerous—than you think.

Phosphorus (P) is also a "primary" soil nutrient. Unlike nitrogen, it cannot be synthesized in a laboratory. Aside from phosphorus-rich manure, it's attained from two key nonrenewable resources: guano and phosphate rock. These resources have been cheaply mined for nearly three hundred years. Geologists now speak of **PEAK PHOSPHORUS**. They predict that by 2030 our demand for phosphorus will outstrip available resources. What country is a major source of global phosphorus? Morocco.

In case you missed the headlines, the phosphorus wars have already begun. Spain relinquished control over the Western Sahara in 1975, after years of United Nations protests. A war immediately followed between neighboring Mauritania and Morocco, with the Polisario Front, a local militia, also joining the fray. Morocco ended up on top and quickly installed the world's largest conveyor belt—more than sixty miles long—to bring phosphate from mines in Bu Craa across the desert to the ocean port of El Aaiún. Increased tensions on the global phosphorus market are inevitable if industrial agriculture continues to depend on these finite reserves.

Josiah Hunt is a **SOIL FERTILITY** specialist who lights fires for a living. I find him in Kauai. Hunt first came to Hawaii for the surfing. Along the way he got a college degree in **AGROECOLOGY** and assumed he would become a landscape designer. Then he read an article about **TERRA PRETA** ("dark soil") and everything changed. The article detailed the work of Johannes Lehmann, a professor of biogeochemistry at Cornell University. "If you're working as a soil scientist in the Central Amazon, you can't help but notice very dark, very fertile soils," Lehmann tells me.

On closer inspection it turned out that these soils were actually modified and improved by Amerindian populations several hundred to several thousand years ago. And the question, of course, is what made these soils fertile for such a long period of time after they had been abandoned.

The answer is something the natives call "terra preta" or **BIOCHAR**. Its porous structure dramatically improves soil fertility by providing a safe haven for fungi, bacteria, and microorganisms. It also retains water, enhances soil aeration, and effectively

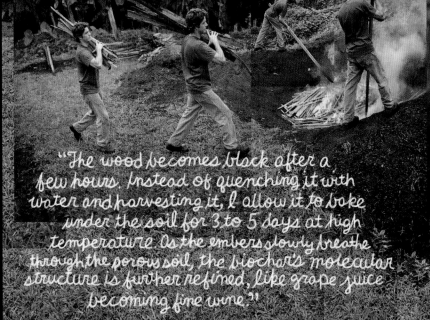

"The wood becomes black after a few hours. Instead of quenching it with water and harvesting it, I allow it to bake under the soil for 3 to 5 days at high temperature. As the embers slowly breathe through the porous soil, the biochar's molecular structure is further refined, like grape juice becoming fine wine."

Josiah's Front Yard
Pahoa, HI
10 December 2012

JOSIAH

THE BIOCHA

someone who specializes in the art and science of prod

JOSIAH TRANSFORMS SCRAP LUMBER FROM A LOCAL SAW MILL INTO A POWERFUL SOIL AMENDMENT INSPIRED E

all the wood has turned to char, cover with dirt for a few days, then moisten and remove. Careful, biochar dries can burn for weeks

room (virtually smoke free) fire. Be sure to use longer pieces of wood first, as they will take longer to burn. 4. When

SEQUESTERS CARBON in the soil. Soil excavations of fertile growing areas throughout the Amazon Basin often reveal layers of this biochar three to six feet deep, but how it's made—and ends up in the soil—remains a mystery.

"We know that simple slash-and-burn agriculture cannot create these terra preta soils," Lehmann explains. "Simply too little biochar would be generated when burning the forest. However, it is plausible that the biomass was charred instead of burned. That could explain the amounts of charcoal found in these soils."

Lehmann's research now follows two paths. The first is to identify *how* biochar is made. The second is to figure out *where* to use it—since biochar has the ability to strengthen otherwise weak soils—which leads me back to Josiah Hunt. After learning about Johannes Lehmann's work, Hunt dug a large hole in his front yard, filled it with scrap wood, lit a fire, waited until the flames reached over his head, then covered everything with dirt and waited. After three days he dug up biochar.

Jeffrey Wallin, a social entrepreneur from Philadelphia, found Josiah soon afterward. Wallin had a plan: build a company—and a machine—to commercially produce biochar. He found a steady source of wood on Kauai, then shipped over an experimental biochar kiln made by an Australian inventor.

Josiah and his assistant, Brett, have been tinkering with the device for weeks. After changing the feed conveyor and tweaking the exhaust system, they're ready. They cut down Albizia trees from a forest only thirty feet away, run the limbs through a wood chipper, shovel these chips (organic material) onto the conveyor that feeds them into the superheated kiln, then wait ten minutes as **PYROLYSIS** turns the wood into biochar. When it works—as it does

CARBON SEQUESTRATION: Excessive carbon in our atmosphere is considered a major contributor to climate change, so practices that remove carbon from the air and capture it in the soil are increasingly in vogue among progressive farmers.

PYROLYSIS: Heating of organic material, such as scrap wood or wood chips, in the absence of oxygen.

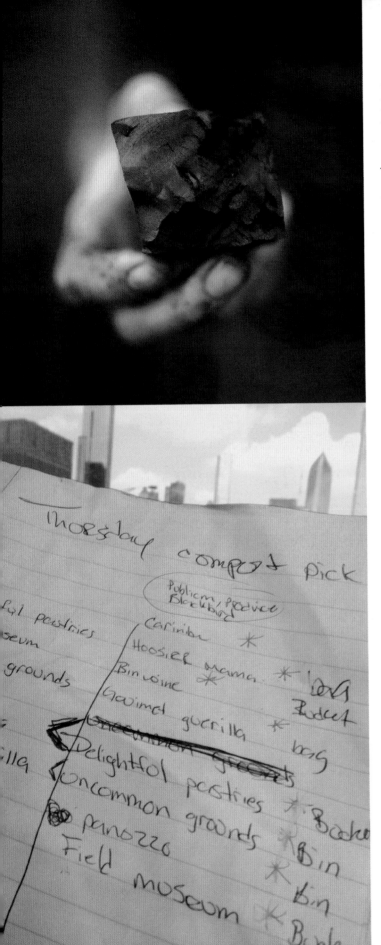

today—it *really* works. Josiah can't shovel fast enough to keep up with the machine. This biochar will soon find its way into gardens and nurseries across Hawaii.

A few days later, I fly to the Big Island and watch Josiah make biochar all over again. His property in Pahoa sits at the edge of a lava flow. Steam sometimes vents from open seams less than two hundred feet away. It's a place that makes you aware of where you are.

Josiah is a collector. His grounds are filled with exotic tropical fruit trees. We spend a few hours wandering through his yard, sampling longans, rambutan, Brazilian plums, Surinam cherries, papaya, pili nut, peach palm, breadfruit, soursop, jackfruit, passion fruit, jabuticaba, and guava (three varieties). The rest of the afternoon is spent tossing wood into an open pit. The fire burns as evening falls. We stand beside it, dusty and tired, beers in hand, mesmerized by the slowly twisting flames until they're finally extinguished by shovelfuls of black dirt.

Jeremy drives me through downtown Chicago in a white panel van. At each stoplight he pulls a clipboard off the dashboard to consult a list of names. These are his morning stops on the **COMPOST CIRCUIT**.

Jeremy works for Growing Power, a Chicago nonprofit that uses a variety of urban agriculture projects to provide self-reliance for low-income communities. They have urban gardens throughout the city. One near the notorious Cabrini-Green public housing project sits directly on top of basketball courts. Two feet of wood chips were dumped on the asphalt, then covered with topsoil, with an irrigation system installed and crops planted afterward; it's a process frequently used to create instant urban farms.

The produce grown here serves some of the city's finest restaurants. I know that because I visit their back alleys with Jeremy. He collects their kitchen scraps, dumps them into plastic garbage cans jammed inside the van, then carts the haul back to the farm. It's August. Hot and humid. By midmorning the stench inside the truck makes conversation impossible, even with the windows down and the AC pumping full blast. The compost will eventually find its way back into the ground, where it will beget another season of fresh produce for those same restaurants as the cycle continues.

THE COMPOST CIRCUIT

to kitchen

to Growing Power's garden

the compost

MIKE
(the chef)

JEREMY
(the composter)

Alley behind
Blackbird Restaurant
Chicago, IL
19 August 2010

THURSDAY'S PICK UPS
1. Delightful Pastries
2. Field Museum
3. Uncommon Grounds
4. Panozzo
5. Hoosier Mama
6. Blackbird
7. Publican
8. Bin Wine
9. Gourmet Guerilla

FARM TO TABLE TO FARM: Jeremy rises each morning at 5:30. After driving Growing Power's staff gardeners to their green projects at Altgeld Gardens and Cabrini Green, he goes on his daily compost circuit, which is comprised of nearly a dozen restaurants in the downtown area which buy Growing Power's produce, then save their kitchen scraps for Jeremy. The compost will break down for nine months then be added to the soil of a Growing Power garden next spring, as the cycle repeats.

BIODYNAMICS

I am given directions to a farm an hour outside Florence, Italy. An old man meets me at the wrought-iron gate. Beyond him stands a dilapidated palazzo, what the Tuscans call *un rustico da ristrutturare*, or a "real fixer-upper," but it's already been here for four hundred years, so I imagine it has a few hundred more left. We exchange pleasantries, then descend to the barn.

I remember the man's massive hands. Like claws. And a back permanently arched. I watch him carry buckets of cow dung into a field, their contents slopping rhythmically back and forth as he makes slow, deliberate strides, his legs clearly struggling with the weight as he crosses the field. His journey ends at a tree stump set beside a rock outcropping. A pile of cow horns waits at the man's feet. Together we scrape the cow dung into the horns. It's slow, messy work, the smell throat-tightening, but that passes. When we finish, he carefully places the cow horns in the bottom of a trench, which he covers with dirt. In spring he'll dig these up again. In strict accordance with the principles of **BIODYNAMICS**, he mixes the dung with water, stirs it in a barrel, then sprays the mixture on his vines. He does this instead of using pesticides and claims it works.

I come back six months later and his vineyard is thriving.

There are farmers who believe in biodiversity instead of monoculture. Farmers who build soil fertility without depending on chemicals. Farmers who go beyond organic. Farmers who do things the old way because they've already seen the new way and said, "No, thank you." Farmers who farm the way their grandfathers did. Farmers who work with nature, not against it. Farmers who treat their land as if it's alive, something to be carefully tended by gentle hands, who observe their fields in silent reverence . . . and learn. These farmers are simply **UNCONVENTIONAL**.

In 1924, ever-decreasing yields lead desperate Polish farmers to try a series of newfangled, largely untested fertilizers produced by German chemical companies. Other farmers remain unconvinced. Maybe they witnessed the effects of these fertilizers' lethal alter egos on battlefields across Europe. "What impact will these chemicals have on our carefully tended land?" they ask themselves. "What happens when you open Pandora's box?"

They turn to an Austrian philosopher named Rudolf Steiner for guidance. He responds with a series of lectures called "The Agricultural Course." The principle—a response to the rapid industrialization of farming—is both holistic and simple. **A healthy farm is a stable, functioning organism, a system with many moving parts—some animal, some vegetable, and some mineral. They are influenced by the rhythms of nature and kept in balance by a farmer who acts as his land's steward and protector.** A key indicator of a farmer's success can be measured by his soil's vitality.

These ideas are soon codified and the biodynamic movement begins. Its adherents number in the millions and produce food you'll even find at your local supermarket, sometimes with a tiny Demeter seal that certifies products as biodynamic, yet you've probably never heard the company's name.

I encounter biodynamic farmers in Caledonia, Illinois. Trust me when I say Caledonia—population 197, down two poor souls from the previous census ten years earlier—is in the absolute middle of nowhere, which makes the discovery of farmer John Peterson, and learning about his parochial biodynamic principles, all the more perplexing.

There are many John Petersons. You can rent the infamous Farmer John documentary that aired on PBS and watch him drive a tractor while wearing a dress, a feather boa wrapped gracefully around his neck. That isn't the John Peterson I meet. My John Peterson pulls up in a fancy new John Deere 6430, a tractor that encloses its driver in a hermetically sealed, air-conditioned glass bubble. Still not seeing it? Imagine crossing a Barcalounger with a jet fighter moving at a snail's pace. That's what we joyride around the farm in while talking about his deeply considered decision to go biodynamic.

To promote soil health and fertility. Farmers closely adhering to these principles are given "Demeter" certification.

they transform into a powerful nutrient and BD Prep known as "500". This is later mixed with water to form a "tea", which is then sprayed on the ground

BIODYNAMICS

IN 1924 RUDOLPH STEINER DEVELOPED THE PRINCIPLES OF BIODYNAMICS TO CHALLENGE THE WIDESPREAD INTRODUCTION OF CHEMICALS IN AGRICULTURE.

He advocated placing manure-filled horns in the ground to "ferment", where they extract silica from the horns as

...quality and regenerative farming practice focused on the integration of plants, animals, soil health and biodiversity. They keep the ecosystem in balance by producing the nutrients needed to nourish all aspects of the farm with a minimum of inputs imported from offsite.

I. COW HORNS
From a small used cattle ranch in Mendocino County

II. COW DUNG
Manure from a local organic lactating female cow

Matt

Jeff fills cow horns with dung (then buries them in the ground)

III. HOLE IN GROUND
Horns go in ground at first "impulse" of Fall and are extracted at first "impulse" of Spring

IV. VINEYARD
Manure is removed from horns and mixed with water. This "Prep 500" is then sprayed in their vineyard in lieu of chemical fertilizers

Philippe

Visiting children from a local Waldorf school (which is based on Rudolph Steiner's principles)

BIODYNAMIC COMPOST

Bringing "aliveness" to the soil and...

Rudolf Steiner first outlined the principles of biodynamics, a system of organic agriculture, in 1924.

He gave recipes for nine preparations. When applied to compost or sprayed directly on soil and plants during the growing year, they stimulate and enhance biological activity on the farm.

Angelic Organics
Caledonia, IL
20 August 2010

THIS COMPOST CONTAINS

Cow manure, straw, a lot of horse manure and vegetable seconds, mixed with the following biodynamic compost preparations:

Prep 502: yarrow supports general growth processes

Prep 503: chamomile guides calcium and potassium processes

Prep 504: stinging nettle enlivens compost with its sensitive intelligence

prep 505: oak bark combats plant "disease" conditions

prep 506: dandelion helps plants access what they need

prep 507: valerian stimulates compost to warmth

The living forces contained by these preps all produce livelihood without harming the realm of the living. They play an important role in the composting process.

"Compost imparts 'livingness' to the soil," he tells me. "It brings the soil to life."

"What about chemical fertilizers?" I ask.

"They instill the soil with an 'imposter vitality,'" he replies. "They're a bit like plastic surgery. They rely on synthetic nitrogen, which provides energy that looks and feels extra real, but it's not sustainable. These chemical fertilizers are like speed for plants. But speed addicts consume themselves . . . and inevitably they die."

Peterson is the only unconventional farmer for miles. He's alone here, but consumers still find him. Today his crew packs CSA boxes for seventeen hundred Chicago families. It would be much easier—and cheaper—for these people to buy produce at the local supermarket, but they don't. Peterson is a trusted face . . . in a trusted place.

Later, we hike across two fields to Peterson's compost pile. **Real farmers don't show you their vine-ripened green zebra tomatoes or that excellent patch of lemon cucumbers. They show you dirt.** Unconventional farming is about building topsoil, and that partly comes from **COMPOSTING**. Every farm has its own recipe. Peterson's includes cow manure, straw, horse manure, plus vegetables that never made it off the farm. To these he adds a series of **BIODYNAMIC COMPOST** preparations, or "preps."

"Preps" are where the mysticism of biodynamics begins. Farmers take their preps seriously. The one called "BD502" is based on yarrow, which helps soil organisms process potassium and sulfur. BD507 uses valerian to break down phosphorus. These recipes—descended from Rudolf Steiner—are closely followed, respected, and unquestioned. When not mixed in compost they're sprayed directly in the fields, not to immediately increase yields but to enhance biological activity in the soil. It's a long-term process.

I stand in a California vineyard one morning, watching a familiar cycle repeat. Cow dung placed inside cow horn. Cow horn buried in the ground, dug up six months later. The dung is removed and mixed with water, then sprayed on a field. It's all part of a complex ritual, one that reinforces the notion of farm as living organism and sets farmers to the task of revitalizing

their soil. I get that. But what exact purpose does the cow dung serve? That's the question I put to Elizabeth Candelario, co-director of Demeter USA, a group that certifies biodynamic practices. And while I don't get a direct answer—it's clearly a concept many folks get stuck on—she attempts to explain the mystical aspects of biodynamics by way of analogy.

She says, "If I just met you and said, 'Oh my gosh, Douglas. You should do yoga because you're going to have this spiritual connection, it's going to open up this pathway inside you,' you might look at me like I was freakin' nuts. But if I came to you and said, 'Look, you should do yoga because it's really good exercise—I mean, you're going to feel better and have increased flexibility,' you might try it. Then, six months from now, you'll come back and describe this totally new connection you're now having between your posture and your breath. I smile, leave it at that, and maybe say, 'That's really cool.' Then two years from now you come back again and go further. Now you talk about your yoga practice. It took you two years, but now you're ready to have a really interesting conversation. It's the same with grasping the principles of biodynamics.

"I think one of the mistakes the biodynamic movement makes, if I can speak about it broadly, is that you have a lot of people who've been practicing for a long time—they're so passionate about it—and they're talking to people who don't have that experience themselves. Something gets lost in translation. My job is to go out and talk about why biodynamics is a really good practice, and I have an inward smile because I know, two years from now, that same farmer will come back and talk to me about a deeper connection he or she is experiencing on the farm."

That "connection" is partly spiritual, but Candelario is reticent to come right out and say it. Instead she talks about what motivated Rudolf Steiner in the first place. "Biodynamics was his answer to the industrialization of farming," she states. "He was saying to farmers, **'Stop looking at your farms as factories, and start looking at them as living organisms—self-contained and self-sustaining—and start following the cycles of nature.'"**

PERMACULTURE

I stand in a respectably sized vegetable garden set in a small clearing between two stands of oaks. A smattering of fruit trees rises in the background. Some cauliflower and Swiss chard are at my left. Artichokes shoot purple flowers into the air a few feet away. My feet rest on a bed of loosely strewn straw that provides cover for rows of fruiting seascape strawberries.

Penny Livingston, a farmer and an educator in Bolinas, California, stands in the same spot, witnessing an entirely different scene. Bird migrations, deer movement, and even raccoons passing along the creek's steep banks define **WILDLIFE CORRIDORS** all around—and above us. These must be carefully preserved. Willows, native currants, and hazelnut trees line the creek. These create natural **RIPARIAN BUFFERS**, earthen barriers that hold the farm's vital soil nutrients—especially nitrogen—in the ground instead of them leaching into nearby waterways. Curves gently cut into the soil create contoured **SWALES**. These capture rainwater, redirecting it into the ground, where it can be utilized by nearby plant roots. Finally, **FOOD FORESTS** (multilayered gardens) allow some crops to be planted above—and below—one another. These carefully considered design principles, each inspired by lessons learned from the natural world, are examples of **PERMACULTURE**.

While biodynamics provides farmers with a vaguely spiritual framework coupled with rigorous practices codified by **DEMETER CERTIFICATION**, permaculture is more observational and free-form; **instead of rules to follow, it offers a lens to look through.** In the natural world, a meadow maintains its equilibrium because all its inhabitants do their part. Bees pollinate. Birds control insect populations and help spread seeds. Plants draw carbon into the soil, then decompose to provide **GREEN MANURE** nutrients for others. The meadow thrives because its "participants" do their share. Nature preaches balance.

"One of the questions we ask in permaculture is, 'How does nature do it?'" Livingston explains. She's a seasoned permaculturalist who spends most of the year traveling the globe, showing farmers how to work with what they have, and to see both their land and ultimately themselves differently. "We look at nature's operating principles, then try to humbly mimic them in our human design," she continues. "We use gravity, the sun, and the landscape's natural flow. We move domestic animals like chickens, goats, horses, or cows around this landscape to keep and maintain meadows. If you do all this right you can actually build soil and sequester carbon in the process."

Livingston shows me recirculating ponds. Dry creek beds brought back to life. Structures built from *cob*, a local mud transformed into pliable architectural clay. Outdoor showers that open onto meadows of chamomile grass. **Nature is clearly trying to tell her something; not only is she listening, she's taking notes.**

Paul Wheaton tells me a cautionary tale about Colorado potato beetles. He's a practicing Montana permaculturalist, and that's no oxymoron. Unconventional agriculture happens everywhere, even Missoula.

"When you have a long row or a big field of potatoes, everything is homogenous. The soil has roughly the same pH. The amount of sunshine the plants get is roughly the same. If one gets hit by the Colorado potato beetle they're all equally susceptible to it.

"In a permaculture system, you don't grow things in rows. You're not growing a big monocrop. You're going to add a lot of texture to the landscape and have your potatoes grow in all kinds of odd places throughout this very bizarre landscape. Then if you see a potato plant that has Colorado potato beetles on it, you don't care. Clearly that potato plant shouldn't be there and Mother Nature, acting through the Colorado potato beetle, is going to take that plant out. Once that potato plant is gone, something else will do really well there. In another patch, there is a healthy thriving potato plant that won't succumb to the Colorado potato beetle."

PERMACULTURE is farming that eschews crop rows and trusts in natural selection. Where I imagine chaos, Wheaton sees a sound investment strategy: Call it the inescapable logic of diversified risk. I need look no farther than my own garden, where gophers got into one raised bed and plowed through two dozen potato plants. In another bed I lost every row of lacinato kale to aphids. I certainly could've benefited from "diversified risk," from spreading things around; that's simple enough with a small garden like mine, but how do these principles work on a larger scale? Will they feed the world?

That question leads me to Geoff Lawton, an international celebrity in the permaculture movement.

"When people talk about rebuilding local food systems, they often refer to scale," I say. "After a certain size is reached you lose 'localness.' Would you say the same thing applies to permaculture? Does it too have limitations of scale?"

"No," he replies. "It's limitless. It's infinite. You just need to re-pattern the system."

"Let's take the state of Iowa, then" I tell him. "Ninety-two percent of the state's agriculturally available land is planted with corn and soybean, yet the entire population of that state is only three million. How would a state like Iowa transition from its current industrial-economic model into something that can be patterned along permaculture principles?"

"By feeding the three million people with two percent of the land that's now presently used for industrial agriculture," he answers.

"And what would happen with the other ninety-eight percent of that land?"

"They would go back to wilderness."

"Wouldn't there be a loss of output?"

"No, there would be an increase in production and people would be happy and healthy, though some rich people involved in large corporations at a distance would be a little unhappy."

"Just so I can wrap my brain fully around this," I continue, "what you're saying is that the entire agricultural output of corn and soy in the state of Iowa could be managed with only two percent of that land and—"

"No, I'm saying corn and soybeans in Iowa are worthless, useless. They only build money for rich people. They don't *feed* people. It actually makes people ill. It destroys the land, creates lots of pollution, and concentrates surplus in the hands of the few. How can that be ethical? It's a destructive, extractive, exploitive process. Instead, we're talking about **BIOMIMICRY**. It increases fertility, increases nutrition, increases health, and creates abundance in surplus. **We only need two percent of our present land area to produce all the world's food needs.** Agriculture in its present industrialized form is absolutely obsolete. We will look back on it as being applied stupidity, rigorous applied stupidity for the sake of lack of understanding."

"Does sustainability play a role in your permaculture work?" I ask.

"Yes, absolutely. A sustainable system produces more energy than it consumes and nothing surplus to maintain and replace its component parts over their lifetime."

"It's almost like a magic show."

"There's nothing magic about the sun," Lawton says. "All energy on earth comes from the sun and all living elements are solar collectors. The biggest transfer of solar energy comes through **PHOTOSYNTHESIS** in the form of starches and base sugars. Those are the building blocks of all life on earth. **If you want surplus, you have to produce more energy than you need to maintain and replace those component parts of the system, to have something extra to trade.** You have to link the sun's energy through photosynthesis to living systems. It's biomimicry and ecosystemic-patterned design. That is the teacher. That is the model. And it is a pretty good one. It's been around since life began on the planet."

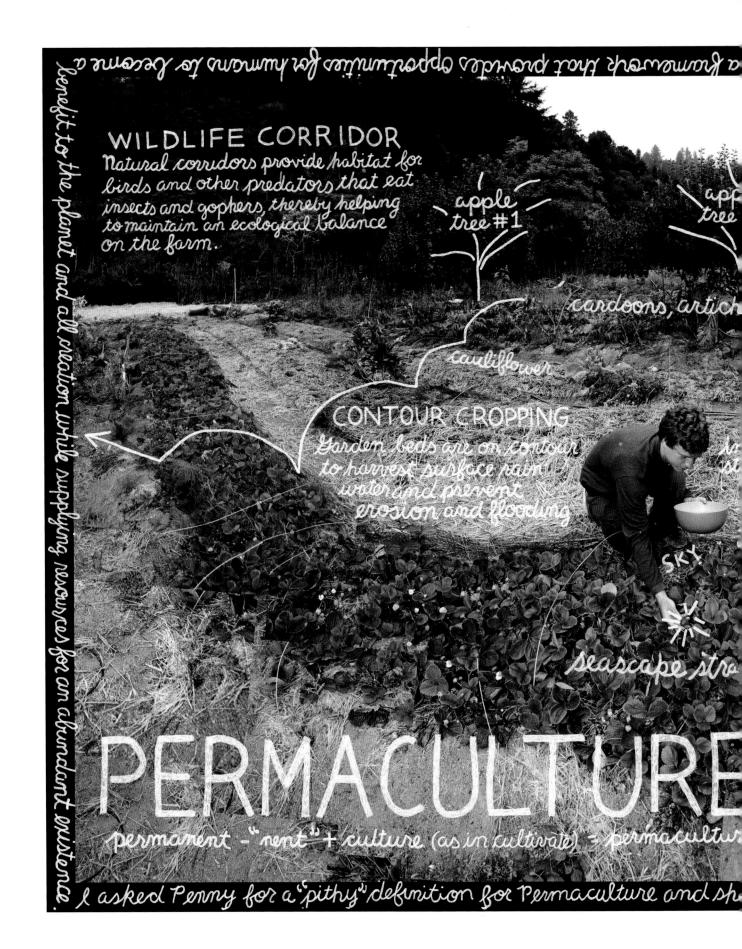

a framework that provides opportunities for humans to become a

benefit to the planet and all creation while supplying resources for an abundant existence

WILDLIFE CORRIDOR
Natural corridors provide habitat for birds and other predators that eat insects and gophers, thereby helping to maintain an ecological balance on the farm.

apple tree #1

apple tree

cardoons, artich

cauliflower

CONTOUR CROPPING
Garden beds are on contour to harvest surface rain water and prevent erosion and flooding

SKY

seascape stra

PERMACULTURE
permanent - "nent" + culture (as in cultivate) = permacultu

I asked Penny for a "pithy" definition for Permaculture and sh

RIPARIAN
BUFFER STRIP

creek

compost

and rhubarb

swiss chard

ED CONTROL
d of herbicides,
s spread between
ows to eliminate
ds which rob
soil of water
utrients.

ies

PENNY

13 September 2009
Regenerative
Design Institute
Bolinas, CA

agriculture, ecology, energy, economy and social justice i

settlements that integrate landscape, water, plants, animals, built structures

d its...a "whole systems" approach to the design of human

SPIN =
SMALL PLOT
INTENSIVE
FARMING

a low-cost, easy-to-learn system that
produces ample high value produce
on less than one acre, making it
possible for first-generation farmers
to get started without a lot of
land or capital investment

Kevin
was a
woodworker
(until he
discovered
dirt).

1) food can be grown closer
to consumers (sometimes
mere blocks away)

2) small amounts of land
can be obtained economically

3) infrastructure
investment is minimal

4) four-legged pests
don't require as much
attention (less deer fencing)

5) cities provide warmer
micro-climates, extending
growing seasons

6) no tractors
needed (just a tiller)

7) little or no outside
labor required

↑
(SPIN FARMING REMOVES 7 BARRIERS
FOR THOSE LOOKING TO START A FARM

Quarter Branch Farm
Lovettsville, VA
16 July 2010

SPIN FARMS INCREASE SOIL FERTILITY
less compaction better aeration
increased levels of organic matter
more available nutrients

In fact, I only became a farmer because of my health. I had ulcerative colitus, but after I switched to fruit and vegetables my health dramatically improved. Buying all that produce cost money, especially the organics, and the quality was poor.

UNCONVENTIONAL

At some point, all progressive farmers were new to their land. Many did their first plantings on fields that had been fallow or recently cleared. They were beginners. They had no well of childhood memories to draw from, no spring plantings, blights, droughts, flooding, price collapses, or meager fall harvests from generations past to serve as cautionary tales. The principles of **SOIL FERTILITY**, of weed and pest control, of optimizing water use, of crop rotations and terroir, were learned over time. These farmers mixed science with observation, then added a healthy dose of common sense. Often they learned from one another. When that failed, they read books.

Every farmer has a favorite. Some have two. Eliot Coleman can't decide between *Make Friends with Your Land* by Leonard Wickenden and *Ley Farming* by Sir R. George Stapledon and William Davies. A self-professed "biblioholic," he carefully tends a library devoted to "alternative" farming practices. These books not only shape how he farms, but also inspire his travels.

"Back in the seventies I read all of [Rudolf] Steiner's stuff and everybody else's," he recalls, "then I organized tours of European organic farms (back when the Europeans were still ahead of us). In the course of the day, we would see three farms. One might be biodynamic. Another would be a **LEMAIRE-BOUCHER** farm, which was a French group that specialized in using coralline seaweed. Or we'd visit a group from Switzerland that never composted anything. They would spread manure before it was more than four or five days old. It was fascinating. In the course of a day you'd see three absolutely beautiful farms. Whether it was the fact that these people paid more attention to how they treated the soil or whether all these systems avoided using toxic chemicals, it made you realize there was more than one way to skin this cat."

Coleman continued to farm, continued to read, and even wrote a few books of his own. Inspiring books, it turns out.

Kevin Grove had two acres in Lovettsville, Virginia, and some ideas. The land was his father's. The ideas were Eliot Coleman's. His books *The New Organic Grower* and *The Winter Harvest Handbook* convinced Grove that he could make a living without having much land or money.

"It made farming seem achievable for an average person like me," Grove tells me. We're standing in his driveway, gazing down at a few rolling acres. It's a front yard that's now a farm. Grove took his life savings—from ten years as a cabinetmaker—and put that into two greenhouses, some basic farming equipment, and a short-row planting strategy that produces two dozen crops for his small CSA and a few regional farmers' markets. Sometimes a kid from down the road helps out; he too wants to be a farmer. This must be what Geoff Lawton's farming 2 percent of the planet looks like.

Grove shows me around. He's noticeably proud of what he's accomplished—much of it alone—and while he's inspired by Coleman's works, he also experiments. He tried P. A. Yeomans's **KEYLINE SYSTEM**, planting raised beds on a contour to capture water (and increase soil fertility). He considered John Jeavons's **BIOINTENSIVE METHOD**, too, even if the thought of double-digging his rows gives him pause. He also thinks the nutrient-dense yields promised by **HIGH-BRIX FARMING** are a possibility, though he admits it would take too much time to create the biologically diverse soil with properly balanced minerals this process requires. Or maybe he'll stay with **SPIN** (small plot intensive farming).

Josiah Hunt and I wind our way up the Mamalahoa Highway north of Hilo. Mauna Kea rises steeply to our left. After passing a dozen waterfalls, even stopping to dip our toes in the water, we pull onto a dirt road that leads to John Caverly's farm. He grows sugarcane—once the dominant island crop but now a local rarity—and papaya, as well as produce for local farmers' markets. We park beside an open shed. It's filled with glass jars, plastic five-gallon containers, pallets loaded with grains of various types. The agricultural version of a medieval apothecary. I hold up a few containers, take a whiff. Molasses. Rice vinegar. Brown sugar. This is what **NATURAL FARMING** looks like, and since Hawaiians are more likely to look east than west, this movement takes its cues from Cho Han Kyu, a Korean agronomist referred to with great affection throughout Asia simply as "Master Cho."

As with biodynamics, Master Cho advocates using a series of preparations. His are based on indigenous microorganisms (IMOs). **He gathers these beneficial microbes from forests, "grows" them out using ingredients found in any kitchen pantry, then adds this mixture directly to fields to boost the soil's microbial life.** Farmers then "feed" these microbes a concoction of fermented plant juice and aerated water, using a process similar to the application of prep 500 in biodynamics. Some say these natural practices are to farming what acupuncture is to Western medicine: an exotic yet highly intimate approach to managing a farm's systemic health. It also

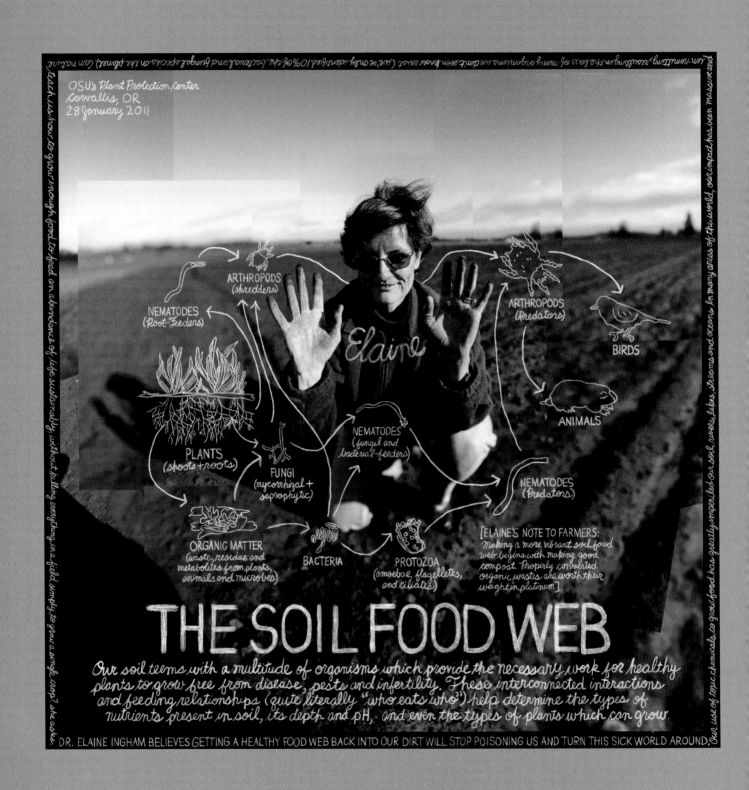

makes economic sense. As Josiah explains, "America has this relationship with organic farming, which is great, but it still depends on dumping bags in a field."

"And that bag has to come all the way to Hawaii by boat?" I ask.

"Yes. The thing about Korean Natural Farming is that you can do it yourself. You're not buying fertilizers. You're making your own. You locally source IMOs and let nature get involved."

This notion of cultivating indigenous microorganisms is unconventional. Dr. Elaine Ingham, chief scientist at the Rodale Institute, would call it strengthening the **SOIL FOOD WEB**. Plastic funnels line the back walls of her lab in Corvallis, Oregon. Each holds a soil sample taken from clients across the globe, all of whom ask the single question: "This is my land, this is my soil, how can I grow what I want to grow?" That answer doesn't come easily. Technicians assess bacterial and fungal biomass, the numbers of flagellates, amoebae, and ciliates, and determine the necessary bacteria-feeding, fungal-feeding predators and root-feeding nematodes. They also look at mycorrhizal colonization and, if required, microarthropod populations. Who knew so much happens inside a clump of dirt?

"What can you do with this data?" I ask.

"If you want to grow weeds, disturb the soil," she answers. "Do not replace organic matter—foods that grow beneficial fungi. Add high amounts of inorganic, soluble nitrate and select for almost strictly bacterial growth, especially anaerobic bacteria in compacted soil."

"I'm not a scientist but it sounds like you're describing the recipe for industrial agriculture," I observe.

Dr. Ingham smiles. It's a Cheshire cat grin, but I don't imagine she'll be disappearing soon.

Her research defines what lives in the soil, and how. She seeks the precise formula, the microbial balance necessary for healthy soils to promote the growth of a particular plant or crop. "Aboveground, people talk about the 'who eats who' food chain," she says. "But in the soil it's not a simple food chain; it's a web of interconnected interactions."

By capturing the complex interplay of microbial activity in healthy soils, Dr. Ingham provides the science that permaculture, biodynamics, and Korean Natural Farming have already intuitively figured out. Chemicals provide transitory benefits, but managing soil requires taking the long view, especially when that soil teems with billions of microorganisms, the very stuff that makes everything you eat possible. We need to be more elegant, more mindful of the soul in Dr. Ingham's soil food web.

BENEFICIALS

The right mix of insects is vital to maintaining a healthy farm. By fertilizing flowers, pollinators increase the productivity of food crops. "Beneficial" predators like ladybugs and beetles consume pest insects as food. **FARMSCAPING** can also help keep pest populations in check by adding plants at field edges and farm margins that create a habitat for **BENEFICIAL INSECTS**. These plants continuously bloom, providing food throughout the year.

Farmers in Eugene, Oregon, tell me about Gwendolyn Ellen's "Farmscaping for Beneficials" project at Oregon State University, so I pay her a visit. We walk the fields behind her research station to inspect **BEETLE BANKS**, rows placed between crops that are "inoculated" with *Pterostichus melanarius*, a predacious ground beetle. They live and raise their families in dirt mounds covered with pollen-rich native bunchgrasses, then venture into the adjacent crop rows to feed. It's an impressive system, one that greatly reduces the use of pesticides.

Misuse of pesticides is one possible explanation for **COLONY COLLAPSE DISORDER**, the disturbing and unexplained disappearance of bees around the globe. When Corky Luster learned of this phenomenon he decided to do his part . . . by making honey. He puts hives wherever people will have them. To see his bees you ring doorbells, pass through gates to enter side yards—being ever mindful of family dogs—sidestep lawn furniture, avoid sprinklers, and duck beneath arbors. Or you go downtown and wave politely at hotel doormen before mounting elevators to the roof. Sometimes you stop beside busy highways. You even visit local farms just outside the city. Luster is an urban beekeeper and his honey tastes of Ballard, an enclave in northwestern Seattle, where most of his bees are found.

Luster's work revolves around the Pacific Northwest's three seasonal **NECTAR FLOWS**, when his bees are most active. First comes the big leaf maple season. That's followed by Himalayan blackberries and then Japanese knotweed, the last two being nonnative invasive species that nonetheless provide bees with a steady nectar supply. When those are out of season, the bees find trails or abandoned lots with lots of blackberries or dandelions. These big-city bees are also productive. While their yields vary from year to year, and depend on weather, Luster often gets 120 pounds or more from a single hive.

I encounter more urban beekeepers at Roberta's, a hipster hangout in Brooklyn, New York. Its heavily graffitied cinder-block exterior leads into an interior adorned with even more cinder block. There are rough-hewn wood-paneled walls, thrift store chandeliers, and tattooed *pizzaioli* making wood-fired pizza. A picture window looks out on shipping containers that house a pirate radio station. Beyond that a profusion of crops grow in recycled barrels and raised beds extending between neighboring buildings. The roofs above this creative mayhem belong to Jon Feldman and his two partners. Their **URBAN APIARY** puzzles me. We stand on a warehouse's tarpaper-lined roof, surveying an industrial vista of concrete, more concrete, and asphalt.

"I always imagined that I'd be in a rural area with a small garden and bees close by," Feldman tells me. "Or perhaps a close friend upstate would allow me to use his land. Never, not even for a split second, did I imagine keeping bees in the city." It was his partner, Brandon, who thought to use a nearby roof.

Back at Roberta's we sample their honey with a few slices of bread. It's not clear and vaguely sweet like the clover honey I have back home. This honey is dark, intense. A bit complex, which begets the question, "Where is the Brooklyn nectar flow?"

Since New York City legalized urban beekeeping, the trend has crossed the five boroughs, but not without complications. Feldman recounts an infamous urban legend that turns out to be true: "The Case of the Mysterious Red Honey." In 2010 local beekeepers opened their hives to discover honey frames stained red. Was it bacteria? A bee toxin? Perhaps bees

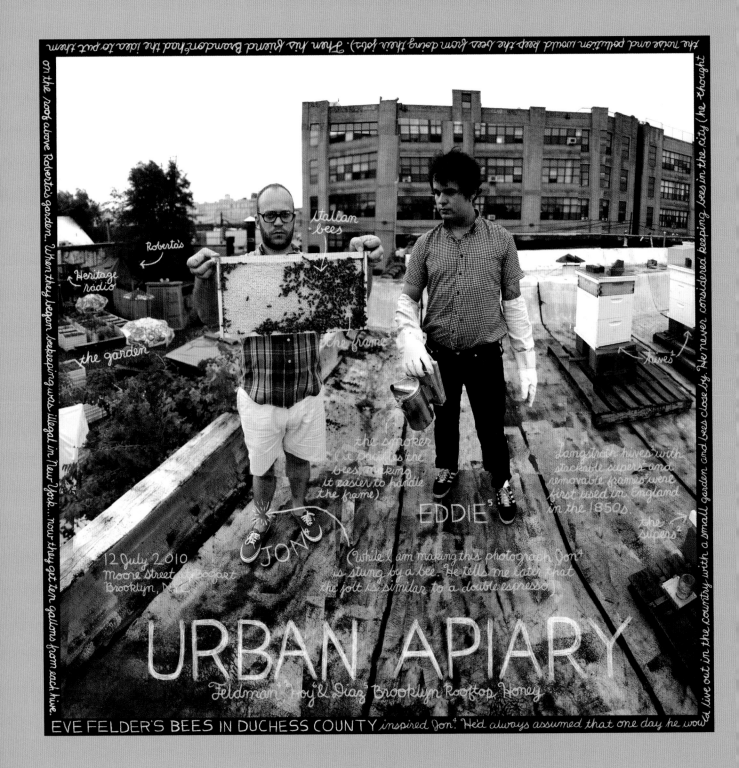

...le by introduction they are sent into the field either in paper bags for ground release or, often being knocked-out at a low temperature (4°C) -

sterile male medflies

Shimon

Egg Laying Cage

BENEFICIAL PREDATORS
Prey upon other organisms and
need many victims to complete
their life cycle. These include bugs,
blady beetles, lace wigs, and
predatory mites.

BENEFICIAL PA
complete their
outside or ins
host. These inc
and parasi

BENEFICIAL MIC
Bacteria, fungi
widely used as b
agents again

BIOLOGICAL PEST

The deliberate and planned introduction of beneficial living organisms to...

THE STERILE INSECT TECHNIQUE (SIT) COMBATS THE SPREAD OF MEDITERRANEAN FRUIT FLIES (MEDFLIES), WHICH THREATEN OVE

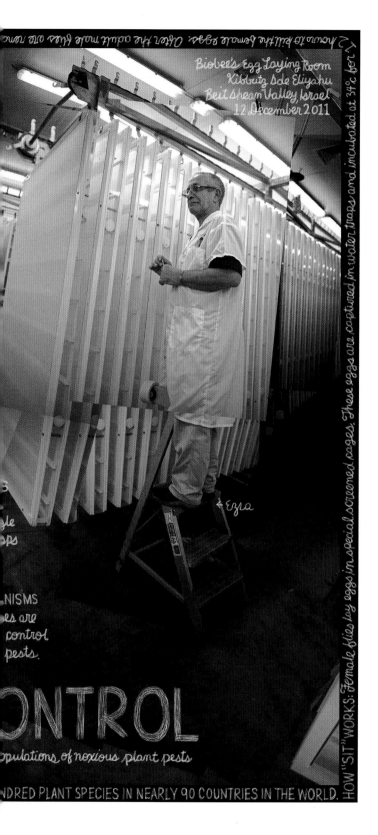

Biobee's Egg Laying Room
Kibbutz Sde Eliyahu
Beit Shean Valley, Israel
12 December 2011

← Ezra

HOW "SIT" WORKS: Female flies lay eggs in special screened cages. These eggs are captured in water traps and incubated at 34°C for 7 hours to kill the female eggs. After the adult male flies are rea...

...NISMS
...es are
...control
...pests.

...ONTROL
...opulations of noxious plant pests
...NDRED PLANT SPECIES IN NEARLY 90 COUNTRIES IN THE WORLD.

pollinating an odd flower like sumac? **For months beekeepers pondered the mystery before discovering its cause: a maraschino cherry company in neighboring Red Hook. Instead of pollinating, the bees had fixated on a steady diet of corn syrup and Red Dye No. 40.** Such are the travails of urban beekeeping.

I discover even more bees at Sde Eliyahu, an Israeli kibbutz in the valley of Beit Shean, near the Jordan River. It's both a working farm and—given the highly entrepreneurial nature of most kibbutzim—a sprawling complex dedicated to the latest scientific innovations. In this case that's **INTEGRATED PEST MANAGEMENT**.

After stepping through a sanitation bath I enter a cavernous warehouse filled with screens. Millions of sterile Mediterranean fruit flies are trapped inside; they eagerly await their shipment to fruit farms around the world that battle ongoing medfly invasions. These flies will mate with fertile females and theoretically end entire medfly bloodlines, thereby saving fruit crops, all without the use of pesticides.

The cool, dimly lit building next door hums, literally. Shelves in every direction are lined with plastic brood nests. Inside each are a bumblebee queen and a half dozen worker bees. In a week these bees will be placed in greenhouses around the world. If industrial agriculture depends upon pollination, and pollination depends upon bees, buildings like these are massive insurance policies against colony collapse disorder.

> **"INTEGRATED PEST MANAGEMENT** is a plant protection regime that utilizes all relevant technologies, in a harmonized way, to sustain pests in a given agro-ecosystem under the economic threshold."
> —Shimon Steinberg, BioBee

APPROPRIATE TECHNOLOGY

As agriculture went industrial and farms supersized, Earl Butz, agriculture secretary to two U.S. presidents in the 1970s, goaded farmers to "get big or get out" and to plant "fence post to fence post." Massive farms require equally massive machinery. But what about small farmers? Who continues to make machinery at their scale? That's what I ask Eliot Coleman, who observes, "A lot of the tools I started out with back in the sixties were really holdovers from the nineteenth century. **No one bothered to make new tools for the small farm because the word was we didn't exist.** We were soon to be replaced by California, so you could ignore us and we would die out. Well, we haven't died out and we still need tools."

Coleman makes many of his own machines. They're what he calls his "projects." One is an alternative to the gasoline-powered tractor. I spy it under wraps at Stone Barns Center, an agricultural teaching farm just a short drive north of Manhattan. The tractor is electric. How does he get these projects made?

"When I come up with an idea for a better tool I look for what I call a 'victim,'" he confides. "For me, the perfect 'victim' is a retired engineer who's played golf for the last six months. He's now bored, but he has a good shop in his basement and he's looking for a project. So I say, 'Hey, have I got one for you.'"

Coleman's electric tractor is an attempt to wean farmers off petroleum, not only in their fertilizers but also in their tractors' gas tanks.

Ian Snider takes things a step further. He's completely ditched the tractor—electric or otherwise—and gone back to **HORSE POWER**, as in power from horses.

I spend a very hot July morning near Boone, North Carolina, with Snider and his wife, Kelly. They're plowing fields where a variety of heirloom and hybrid organic fruits, vegetables, and root crops grow. I even find Honey Drip sorghum cane. From that they'll make sweet sorghum syrup, a traditional sweetener in southern Appalachia.

"The closer we get to the smaller o

DOUG MOSEL (and combine) lend a hand on his friend's nearby farm.

HEGE 140 COMBINE
"Research" or "small plot" combines like this are expensive, highly specialized and manufactured in relatively small numbers. Usually used by universities or seed developers to harvest v small test plots, they are not intended for productie

APPROPRIATE T
Environmentally sound technology designed to of a specific geographic area while ideall

AS FARMERS RE-INTRODUCE WHEAT TO STRENGTHEN TH

HORSES = SOLAR-FUELED +
SELF-REPAIRING ANIMALS +
(SOMEWHAT) RENEWING +

KELLY

Ian and Kelly Snider
Yates Family Farm
Fleetwood (just east of Boone), NC
3 July 2017

FRANK

IAN

HORSE POWERED

Working with equines in a symbiotic agricultural relationship
(they complete tasks which require great effort, while sustaining themselves and us from the land)

I ASK IAN, "WHY HORSES?" HE SAYS, "THIS CHOICE IS NOT A NOSTALGIC ONE, OR ONE MOTIVATED BY OBSTINATE ANTI-ESTABLISHMENT SENTIMENT.

Frank is at the helm. He's their fourteen-year-old Suffolk cross gelding, what Snider calls "a real good dude." Frank runs on "farm fuel"—grass, grain, and water—which makes him cheaper than his gas-powered cousins.

"Our horses are an **APPROPRIATE TECHNOLOGY** in that they are an element of a system that is simultaneously a source of power for development as well as a source of beneficial sustenance," Snider explains. "They perform heavy work, heal from breakdowns mostly on their own, run on solar-charged biomass, produce nitrogen-rich fertilizer, and replace themselves along the way. When it comes to farming that is about as appropriate as it gets. They are also very quiet and smell good."

What tractor can make such bold claims?

Snider admits that working with animals puts them out of step with the current trend of modern agriculture, but at least they've avoided the most common hazard facing young farmers: going into debt to buy farming equipment (though he does possess a portable James Bondesque tack room jammed with carefully braided reins and bridles, padded leather collars and harnesses, polished bits, and buckled girths).

He's also emphatic that his choice is not nostalgic. He's not living out an agrarian fantasy set in the eighteenth century. Nor is he political. He has no antigovernment statement to make. Don't call him an "antitechnologist" either, or a "Luddite," though he does like reminding me that **"even today, in this post-postmodern world of digital everything, our machines are still measured with the term horsepower."** And while I never ask about his specific religion, he assures me it's not about that, either. Not exactly. It's spiritual, but on a personal level. The couple claim to be motivated by a "deep obligation and sense of responsibility for those who will come after us. By utilizing the culture of the past we can find the most hopeful path for the future."

In Northern California, Peter Buckley decides that the world might not need another bottle of Pinot but it could use some local grain. It's nearly July and the barley at Buckley's Front Porch Farm in Healdsburg is ready to harvest. Doug Mosel glares down at me from atop his recalcitrant Hege 140. At this time of year he transports this small German combine to farms across Sonoma and Mendocino counties, but today the beast stands idle at the end of a crop row, stubbornly refusing to cooperate. Inspection plates are open. Tools are spread about. The Hege is a delicate machine, one better suited for seed research plots at land grant universities than the rigors of agricultural production.

This combine expresses both the potential for and limits to appropriate technology. As farmers return

different crops to local food production, grain remains a daunting challenge. Most small farms simply don't have the right tools.

"Small-scale combines are very expensive," Mosel points out. "They're specialized and manufactured in relatively small numbers. Farmers without lots of money have to search for affordable, older equipment, which is difficult to find."

The search for appropriate technology doesn't end with finding the right tractor or thresher, either. Instead, it creates a cascade effect. Next comes finding appropriate separating equipment as well as facilities willing to clean small quantities of grain. Our agricultural system has been very good at getting big. Going small will take even more work.

"We like working at a human pace," Ian Snider observes. "Machines almost always dictate that you move faster in order to keep up. **Horses plow, or do any work for that matter, at a pace that fits the human body's more natural rhythms.** We both need to breathe steadily. We need to take a moment to check our footing, react to danger, preparing ourselves to begin or end a task, even take a break. We are able to wait for each other. Focusing on the details of soil, tilth, measure of stride, the sounds of leather, metal and rock, tone of voice, birds singing, tone of respiration, wind blowing, shifting of contours on the ground, minuscule twitches of the driving lines, balancing the plow, watching the ears, and even feeling the earth roll in waves beneath you. It all requires a presence of mind, body, and soul rarely required in life. Man and beast, existing in a moment of grace."

"APPROPRIATE TECHNOLOGY is a holistic approach to human problem solving. It involves working with nature—of which we are a part—to develop and sustain systems beneficial to the health of the whole."
—Ian Snider, Yates Family Farm

THE OSTTIROLER (from Austria)
When industrially milled at high RPMs, the wheat's components (endosperm, germ, and bran) rapidly separate. But the Osttiroler's slower turning mill stone allows oil present in the germ to rub into and perfume the endosperm (white part of wheat). This retains the wheat's native fragrance.

Farm and Sparrow
Candler, NC
1 July 2011

the hopper

David

Turkey Red Winter Wheat

LANDRACE WHEAT
a genetic grandfather wheat that has been introduced around the world and given birth to new geographically-specific varieties wherever it has landed

John McEntire's
Turkey Red Winter Wheat
Peaceful Valley Farm in Old Fort, NC

THE DIFFERENCE BETWEEN MODERN AND HEIRLOOM WHEAT? THAT'S A GOOD QUESTION. Modern wheats are bred to produce high yields

modern and ancient wheats," David says. "Modern ones feed the world. Ancient ones provide valuable breeding information to make agriculture more sustainable."

but only when given high doses of artificial fertility dependant on fossil fuels. Heirloom wheats have a highly adaptive genetic database. Landrace wheat varieties, for example, are hardier and migrate across the globe. One variety, Turkey Red Winter Wheat, can manifest over forty iterations depending on environmental conditions. "We need both

IDENTITY-PRESERVED GRAINS

David Bauer lives in Candler, a remote valley community ten miles west of Asheville, North Carolina. Fruit trees and a small vegetable garden greet you amicably as you pull up to a modest country home. The attached garage holds no cars. These have been replaced by a massive wood-burning brick oven. Bakers mix dough, shape loaves, split wood, mill flour, feed chickens, stoke fires, and bag bread in no particular order. Farm & Sparrow is a "diversified" rural bakery whose "completely from-scratch holistic approach" also includes using flour grown by local farmers.

Back when wheat was local, the area boasted thirty stone mills. Now only two remain. Local wheat. Local mill. Local baker. These nested relationships disappeared with the industrialization of our food system, but in many communities across the country local grain producers have returned. They're discovering (or rediscovering) what used to grow in their communities.

The area around Asheville shifted from wheat production to tobacco more than seventy-five years ago. Convincing these farmers to go back to wheat is challenging, but Bauer is lucky.

"When I met John McEntire I found someone who shared the same values I had," he explains. "He was willing to take the risk and grow turkey wheat and Abruzzi rye, and to work with us directly on a small scale. That was a really big deal because in these mountains farmers aren't set up for grain growing."

Being "set up" means having the **APPROPRIATE TECHNOLOGY**. McEntire, an experienced family farmer, has both a forty-inch Allis-Chalmers combine and a larger self-propelled Gleaner combine harvester at his Peaceful Valley Farm. He also possesses the requisite mechanical skills to sustain such an antique machine-dependent undertaking (our first meeting takes place while he works on a Chattanooga No. 44 sorghum mill, its rusted innards spread like clock pieces across his workbench). Lastly, he has the right infrastructure in place, with most of it already dedicated to his heirloom corn operation.

After McEntire grew out Bauer's old varieties, other farmers followed.

"His taking that risk and working with us led to those grains becoming established across a number of states," Bauer points out. "Now, if John has a bad year, we don't lose the grain. We have someone else to turn to because of his work in multiplying out the seeds and giving them to other farmers."

Before each season, Bauer sits down with local growers. He proposes specific grains, and they say how many acres they plan to seed.

"To make the grain-growing relationship work between farms and bakeries it needs to be collaborative," Bauer says. "It can't be one bakery and one farm, one bakery and two farms; that puts too much pressure on each person. If a farmer decides to grow four acres of a certain wheat, and we can only purchase two, I can always call another bakery and negotiate for them to purchase the rest."

These discussions about who plants what are ongoing; they happen throughout the year.

Healdsburg

Front Porch Farm grows a variety of wheats, including Frassinetto (Triticum aestivum), for local food producers.

CALIFORNIA

IDENTITY

Labeling that protects the chara... as enhanced by specific place...

BOB KLEIN OF COMMUNITY GRAINS WONDERS IF COMMERCIAL MILLERS CAN TAKE A WHEAT SEED APA...

Steve Jones, a plant breeder at Washington State University, claims the origin of this "small grain" movement can be traced to growers eager to "kick the commodity habit" and abandon an industry where grain prices are set in faraway places like Chicago, Kansas City, and Minneapolis.

"We now have craft bakers and wood-fired ovens popping up all over the country," Jones says. **"They don't want the same flour that bakeries use in grocery stores. They want better flour. They want local flour. They want flour that has a name and a face and a region. And they want to know that they're helping their local growers."**

Bob Klein, a Berkeley, California, food entrepreneur, has a name for wheat with "a name, a face and a region": **IDENTITY PRESERVED**. Each package of his

152

Frassinetto

Originally developed near the town of Frassinetto, Italy in the 1920's. Reputed to be an excellent variety for making pasta or bread.

ITALY

PRESERVED

istics which define a unique grain or crop, farming practices and milling processes

the seed's three components germ, bran, and endosperm, then reconstitute them later. We all added

O PUT IT BACK TOGETHER AGAIN TO MAKE WHOLE GRAIN FLOUR. He says, "Steel roller mills separate

Community Grains wheat pasta and flour comes with a tag that states the wheat's provenance and answers the questions, "Who developed the variety and when? Who farmed it? When was it planted and on how many acres? When was it harvested? How was it milled? The package tells you it didn't come from Timbuktu sixteen years ago or sit in a warehouse off the coast of Guam. It tells you what it is."

I later visit a small farm growing Bob Klein's identity-preserved grains, then return home to make their pasta. It's an unexpectedly powerful experience, knowing that a product I always associated with amber waves of anonymous grain can suddenly become local. This wheat has a name: hard amber durum. A face (actually two): Matt Taylor's and Peter Buckley's. And lastly a place: Front Porch Farm in Healdsburg, California.

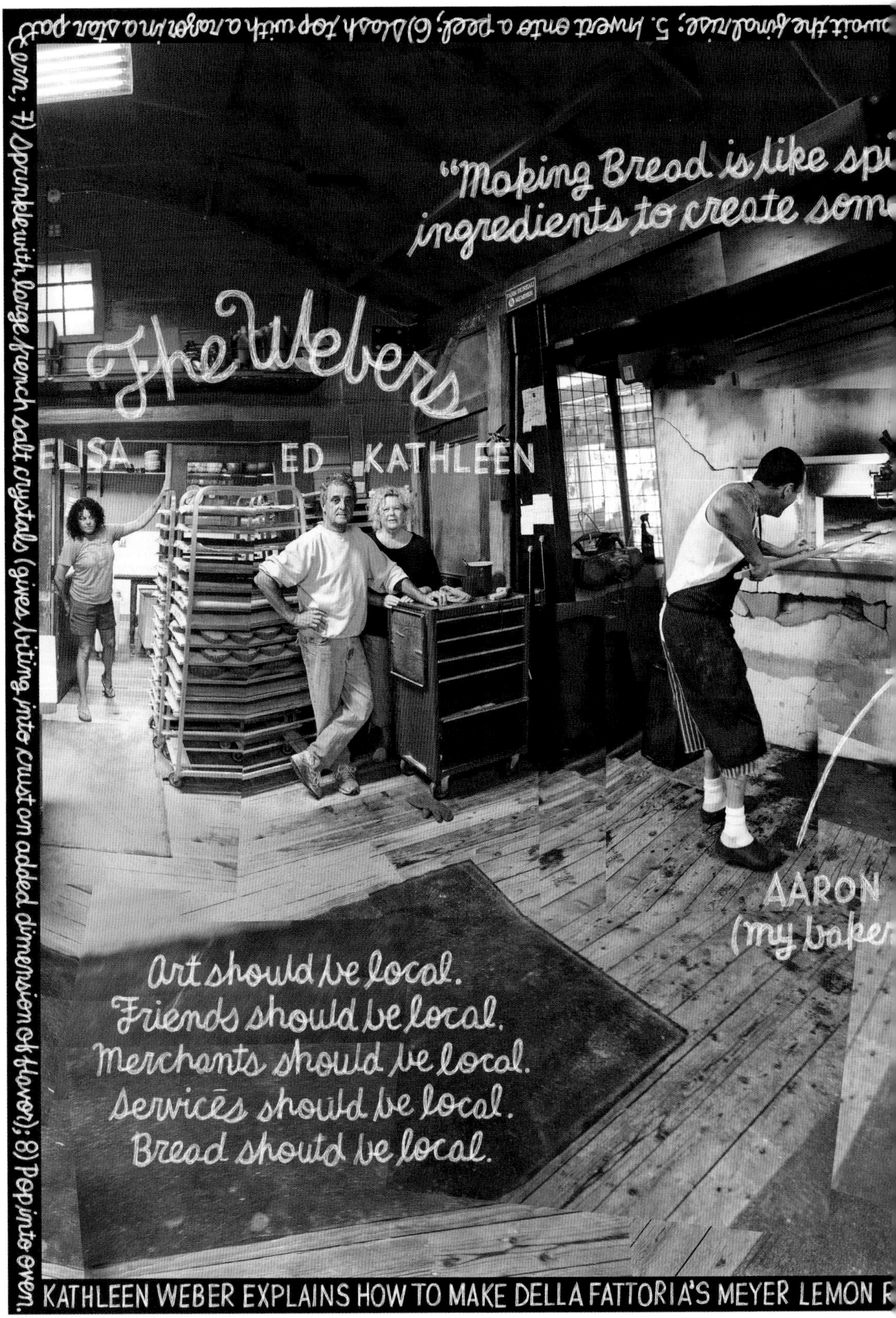
await the final rise; 5. Invert onto a peel; 6) Slash top with a razor in a star pat
tern; 7) Sprinkle with large French salt crystals (gives biting into crust an added dimension of flavor); 8) Pop into oven.
“Making Bread is like spi
ingredients to create som
The Webers
ELISA
ED
KATHLEEN
AARON
(my baker
Art should be local.
Friends should be local.
Merchants should be local.
Services should be local.
Bread should be local.
KATHLEEN WEBER EXPLAINS HOW TO MAKE DELLA FATTORIA'S MEYER LEMON R

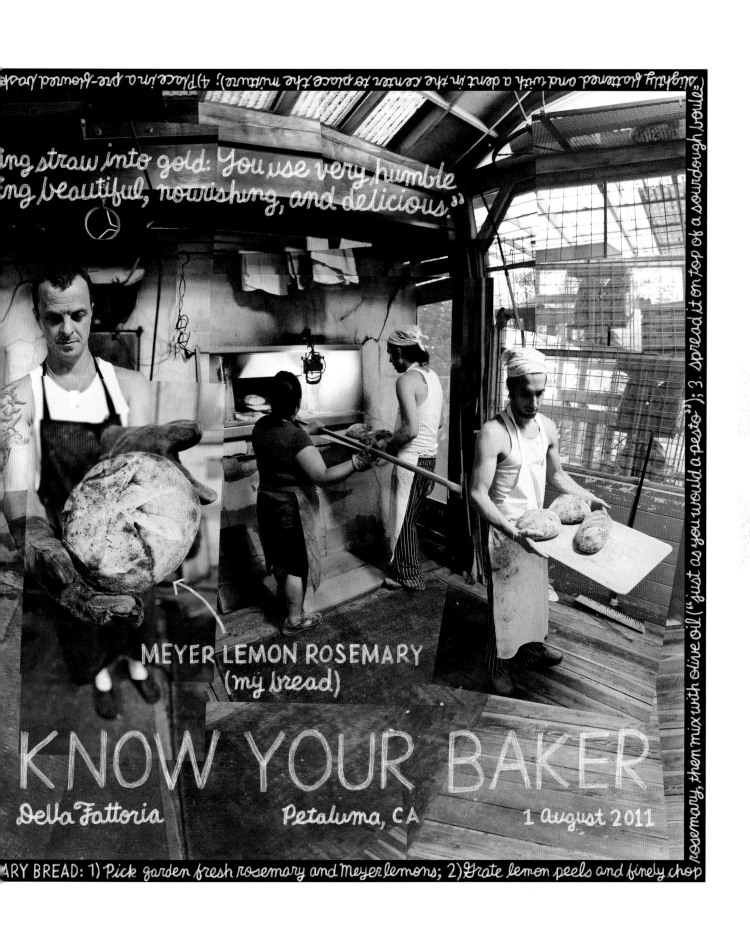

MEYER LEMON ROSEMARY
(my bread)

KNOW YOUR BAKER

Della Fattoria   Petaluma, CA   1 August 2011

experience this again in Oregon. My morning starts at a chamomile harvest outside Eugene, and I need to be in Portland for dinner. A slight detour lands me at Open Oak Farm; I meet Sarah, Andrew, and Cooper in their dusty front yard. Seed-sifting screens are stacked by a tree. A box fan used for cleaning seed is visible inside the garage. We inspect a few greenhouses just beyond the tree line. The test plots inside display farming experiments, ideas in motion; they're like an artist's sketchpad, except these works grow in the ground. A mad profusion of greenery pushes up to the ceiling.

This is what young farmers look like today. For inspiration, they gaze as much backward as forward. They're overeducated and hyperaware. Big on ideals. Short on cash. Driven not only by practicality—they have to make a living, after all—but also by the conviction that agriculture is restorative; it can bring nature back from the brink.

"We wanted to start farming to get away from the meaningless busywork of most occupations," Andrew explains, "and do something real that has a substantial impact on us and our environment. Eating is one thing we all do and agriculture has the largest impact of any endeavor on the planet. Therefore, helping to create a better food system is possibly the most powerful thing we can do to turn this human catastrophe around."

Their plan for building a better food system waits on the other side of a small wood bridge. Beyond that a combine rusts in an open field. They'll soon coax it into action, as a dozen **HERITAGE GRAINS** grown on this small property are ready to harvest. A few of these come with a story. William J. Sando oversaw a USDA wheat-breeding program back in the 1920s. Two promising varieties emerged: SS791 and SS176. Ninety years later these seeds found their way into the hands of Tim Peters, an independent plant breeder, who gifted them to Open Oak Farm. It's their first season in the ground, but Andrew is already optimistic. They're disease-resistant and have proven highly adaptable to the farm's organic growing practices. Andrew also thinks they can be a "great start to establishing a local grain **TERROIR**."

156

COPPER

ANDREW

SARAH

"SS791"
An annual wheat first bred
by William Sando 80 years ago.
The original crosses were
made from 1923 to 1935:

Triticum durum (annual wheat) +
Agropyron intermedium (perennial wheatgrass

SS791
1) very disease-
resistant and tall
2) extremely weed
competitive
3) ideal for organic
growing conditions

ly
vers
d it."

E WHEAT

ken human-plant co-evolution, effort, and reverence

Adaptive Seeds
Open Oak Farm
Sweet Home, OR
24 August 2011

EMPHASIZED ASPECT OF OUR RE-LOCALIZED NEO-DECENTRALIZED FOOD SYSTEM." – ANDREW STILL

I take two planes, rent a car, and drive three hours— taking a most agreeable pit stop at Bobo's in Topeka for a chocolate malt—before arriving at the Land Institute in Salina, Kansas, to meet Wes Jackson. **"Got tired of taking your pictures of tomatoes?" he asks. "Vegetables are only three percent of American agriculture. People eat wheat, a lot of wheat, and all you California people talk about are organic tomatoes."** He's not joking. Jackson is all about wheat, but not the kind you expect. His inspiration came from the Konza Prairie near Manhattan, Kansas.

"It's a native, tallgrass prairie that was never plowed," he explains. "It was wonderful. Here was the prairie, which didn't have soil erosion beyond natural replacement levels, didn't have any fossil fuels it was dependent upon, and with species diversity there was chemical diversity.

"While we stood in that prairie, all around we could see corn, wheat, sorghum, sunflower, and soybean fields. What a tremendous contrast. They had soil erosion. They were fossil fuel dependent. It became clear that the distinction had to do with the one system—nature's system featuring **PERENNIALS** grown in mixtures—versus a grain agriculture that features annuals largely grown in monocultures. So that's what set me off. We began the journey to see if we could build an agriculture based on the way natural ecosystems work."

American agricultural production is based on the use of ANNUALS. Fields are plowed. Crops are planted, watered, fertilized, then harvested. The cycle repeats. It's an inefficient, energy-intensive process, one Jackson hopes to change. "An annual is a plant that requires you to tear the ground up or treat the ground with an herbicide to get the grain to germinate," he continues, "so nature must be subdued or ignored to make the plant grow. On the other hand, a perennial just keeps coming up every year. The top parts may die back, but next year they come up on their own."

The Land Institute is developing a "plant once, harvest for years to come" **PERENNIAL WHEAT** for commercial use. Its potential impact on American agriculture could be huge. As these perennials grow, their roots venture deep into the soil, allowing them to mine nutrients and water that would otherwise be unavailable to an annual plant.

Jackson and I examine a series of promising perennial wheat varieties at test plots scattered around the institute. Some grow inside plastic pipes. A tractor lifts one out of the ground to reveal the plant's extensive root system. We see more roots inside a nearby laboratory. They're carefully placed on white tables, and handled delicately, like rare examples of sixteenth-century Victorian lace. They're also ten feet long. The annual wheat root systems placed beside them look sickly in comparison.

The Land Institute's quest for commercially viable perennial wheat is a long-term project, with positive results not expected for decades, but Jackson is an effective manager of impatient expectations. He doesn't want to simply erase industrial agriculture and return to the past. That's not an option. "I don't want the agriculture we had in 1920," he says. "Those are annual systems. I don't want the agriculture we had in 1880. I don't want the agriculture we had in the time of the Romans or the Greeks or the early Hebrews. I don't want the agriculture we've had for ten thousand years. We have a chance for a different kind of agriculture now. It uses the ecosystem concept instead of nature being subdued or ignored in order to get annual seeds to sprout: 1920 agriculture was destructive; 1880 agriculture was destructive. George Washington and Thomas Jefferson both destroyed soils. We're talking about a new paradigm here."

The Land Institute
Salina, Kansas
24 August 2010

PERENNIAL VS. ANNUAL

perennial = plant once   annual = replant every year

PERENNIALS HAVE
LARGER ROOT SYSTEMS
1. improved soil stability = less
   need for tillage + reduced erosion
2. reduced fossil fuel consumption
3. better managed nitrogen
4. reduced need for pesticides
5. less labor intensive
6. increased soil water storage
7. better carbon fixing
8. greater biodiversity

the dehuller

JOHN

WES

EMILY

four feet long

annual wheat
(Triticum aestivum)

ten feet long

intermediate wheatgrass
(Thinopyrum intermedium)

WES JACKSON HAS BEEN DOMESTICATING INTERMEDIATE WHEAT GRASS AT THE LAND INSTITUTE SINCE THE 1970s.

ASPHALT AGRICULTURE

It's 7 a.m. and Novella Carpenter is milking a pygmy goat in the kitchen of her apartment in West Oakland, California. It's in a low-income "lock your car and look over your shoulder" neighborhood at the intersection of three freeways and a transit line. In fact, I can see Interstate 980 from her kitchen window, along with an oversize vegetable garden on the adjacent corner lot. Carpenter calls it Ghost Town Farm.

She planted the garden two weeks after moving in.

"I went around like any do-gooder and knocked on people's doors," she says. "I was like, 'Hey, let's start a garden, you guys!' And people were like, 'Yeah, we get our groceries at Safeway, no thanks.' So I figured out I had to do it myself."

Which she did. The lot's dominant architectural feature was a concrete foundation left behind after an apartment building was demolished. She buried that under dirt and a variety of scavenged soil amendments. Then she planted a garden. The neighbors came soon after.

"If you really want to know your neighbors, start a garden in the middle of where everyone is," Carpenter says laughing, "because they're all going to come out and ask you questions and give their advice for growing stuff."

The garden flourished. Bee boxes followed, as did chickens, turkeys, and pigs. She placed her husk of rabbits on an upstairs balcony overlooking the street.

"The gate's open," she continues. "People come in. They harvest food when they want it. For some reason, picking it themselves makes it more like 'We *own* this. This is a community garden,' and I say, 'Whatever. Fine.'"

Carpenter was motivated to start an urban garden out of a desire to know not only where her food came from, but how to grow it herself. **URBAN AGRICULTURE** isn't about to feed entire cities, but it will make these people more aware of what food is, which is important because according to the 2010 census, 80 percent of Americans live in or near cities.

"The idea of local eating is really beautiful," Carpenter observes. "It's something I can get behind. You support your local farmer and have this . . . It's almost like you get to eat their view, their beautiful patch of rural landscape. But the reality is—in places like my neighborhood—they'll never go to that rural place. They'll never be able to afford local, organic, biodynamic, sustainably raised food. So that's why I think growing food in the city is so important, because you're localizing the food in the city where people live."

Eli Zigas works at SPUR, a San Francisco think tank for urban planners. He's paid to ponder the complexities of embedding urban agriculture within local food systems.

"I think urban agriculture's real contribution is not in the quantity of food it grows," he states, "but the number of people it touches who can then understand and learn about food, how we grow it and how it feeds us."

When Zigas says "how it feeds us," he doesn't just mean nutritionally. He also means how food feeds our minds and our souls. Urban agriculture doesn't need to replace a rural food system, but growing and harvesting closer to where people live gives them access to fresher, more nutritious food.

"If you only have a global food system, you lose the farmland around most major cities, which leads to more sprawl," Zigas observes. "People lose the sense of where their food's coming from—that ecological awareness— which is important to have." What does Zigas advocate at a city's **URBAN EDGE**? Greenbelts.

"As a planning mechanism, a **GREENBELT** of agriculture around cities helps focus density," Zigas notes, "and depending on how the food is transported, it helps create a less energy-intensive means of distributing produce."

To witness Ben Flanner's inspired contribution to New York City's greenbelt, you take an elevator and push the "R" button. Get off on the top floor, follow the signs up two flights of stairs, swing open a heavy plated door, and behold the largest **ROOFTOP FARM** in the world.

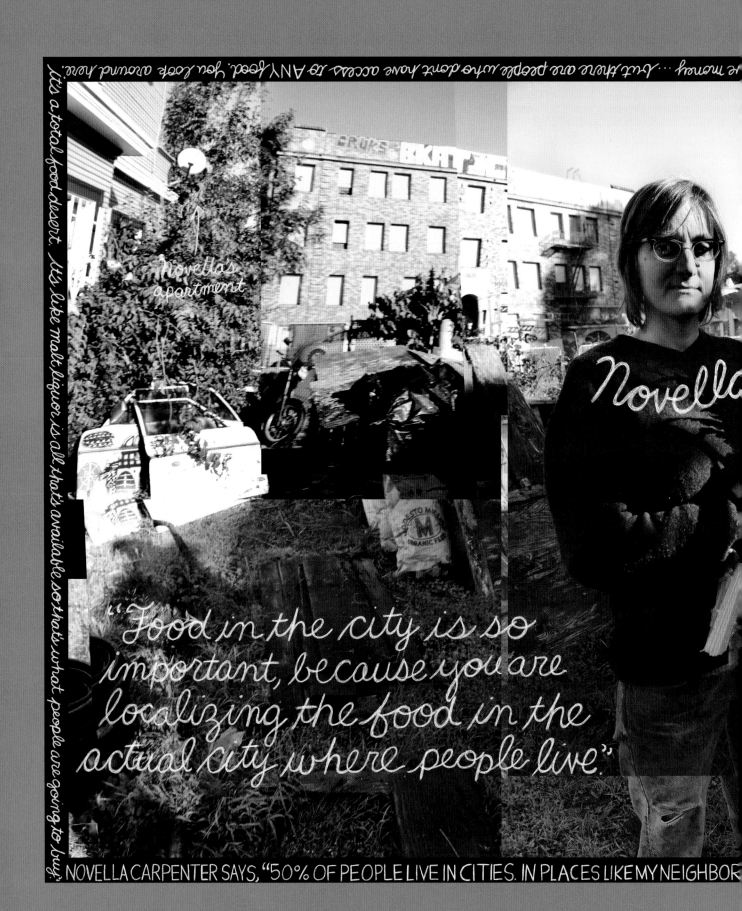

re money...but there are people who dont have access to ANY food, you look around here...

it's a total food desert. It's like malt liquor is all that's available so that's what people are going to buy

Novella's apartment

Novella

"Food in the city is so important, because you are localizing the food in the actual city where people live."

NOVELLA CARPENTER SAYS, "50% OF PEOPLE LIVE IN CITIES. IN PLACES LIKE MY NEIGHBOR

My buying local food, but the thing is, if you look at our society, that's great for people w...

cardoon

Ghost Town Farm
Oakland, CA
9 December 2010

URBAN FARMER

(because, that's what she is)

urban = a city like Oakland

farmer = one who raises food for others

Her mission is to grow food for herself, her friends, her neighbors, and the larger community.

We live in a society where supposedly you can vote with your fork, and you can change the system just

THEY'RE NEVER GOING TO BE ABLE TO AFFORD LOCAL, ORGANIC, SUSTAINABLY RAISED FOOD.

A brigade of zealous young workers—idealistic, heavily tanned, dressed in shorts and bright T-shirts—are living the Brooklyn foodie dream: They're farming a one-acre plot set on a city roof, with the Manhattan skyline serving as a shimmering mirage across the nearby East River, and picking a variety of salad greens and kale. Lots of kale. The whole place is Flanner's idea, and it's no trivial undertaking. You don't just shovel a million pounds of topsoil onto the roof then start planting tomatoes.

First, a barrier is rolled out to prevent roots from penetrating the roof's surface. A felt absorption layer comes next. This captures excessive moisture from heavy rainstorms, then releases it back to the soil and plants when it's needed during dry conditions. A protective layer is placed above that to protect the absorption layer from soil contamination. As for the soil, it's actually a blend of organic compost materials and lightweight, porous stones. These are designed to break down and add the trace minerals vegetables need.

Flanner's customers include local restaurants, farmstands, farmers' markets, and a fiercely local CSA that wants its urban produce from within a five-mile radius.

Colin McCrate builds urban farms in Seattle. His projects range from front-yard farms to epicurean rooftop **FOOD JUNGLES** for local businesses. I follow him through a busy restaurant. **As we pass the kitchen, workers stop. To them McCrate's a familiar, albeit perplexing character: the man who grows vegetables on the roof.** We continue down a hallway, turn a corner, pass the back bar, then climb steep wooden steps to the tarpapered roof. His garden, which consists of wood-framed raised-bed planters and plastic baby pools converted for salad duty, occupies the open space between ceiling exhaust fans and boxy air-conditioning units. The summer weather in Seattle is always a challenge—rain, more rain, followed by unpredictable periods of pounding sun—so McCrate focuses on short-season crops that deliver high yields in small spaces. It's all part of what McCrate calls **HYPERLOCAL FOOD PRODUCTION**—food that doesn't travel farther than a kitchen one flight of wood stairs below.

Other urban gardeners are less demanding. They ply their trade wherever they find dirt and a working sprinkler system. In Los Angeles, I make an appointment to meet a **GUERRILLA GARDENER** at the base of a Hollywood freeway on-ramp. His *nom de bloom* is appropriately "Mr. Stamen."

A kid with a small shovel in one hand rides up on a mysterious black bicycle, a bandana strategically positioned to obscure his face. He gracefully dismounts, sets his bike against a palm tree, then directs me to his "garden": a bunch of effusive rosemary bushes growing from the asphalt sidewalk just off Sunset Boulevard. He and his associates have planted herbs in two dozen similarly impromptu garden sites across the city, though he's quick to admit that most dirt at these locations is better suited to hard succulents and drought-tolerant native perennials, which they also plant.

"**URBAN AGRICULTURE** is about growing and distributing food in the communities where people live, work, shop, and play. If you are out in your field or in your garden and you stand up and look around and there are neighbors and people around you who have some kind of influence over what you do there, you're an urban farmer."
—Katherine Kelly, cofounder and executive director of Cultivate Kansas City

"An **URBAN EDGE** is the border between a city or suburb and its surrounding environment, often defined by the boundary of urban infrastructure, such as sewers, or by a sharp contrast in density or the built environment."
—Eli Zigas, SPUR

"A **GREENBELT** is a ring of land encircling an urban area free of residential, commercial, or industrial development. Often protected by policy, a greenbelt usually consists of open space, forests, parkland, and/or agricultural land."
—Eli Zigas, SPUR

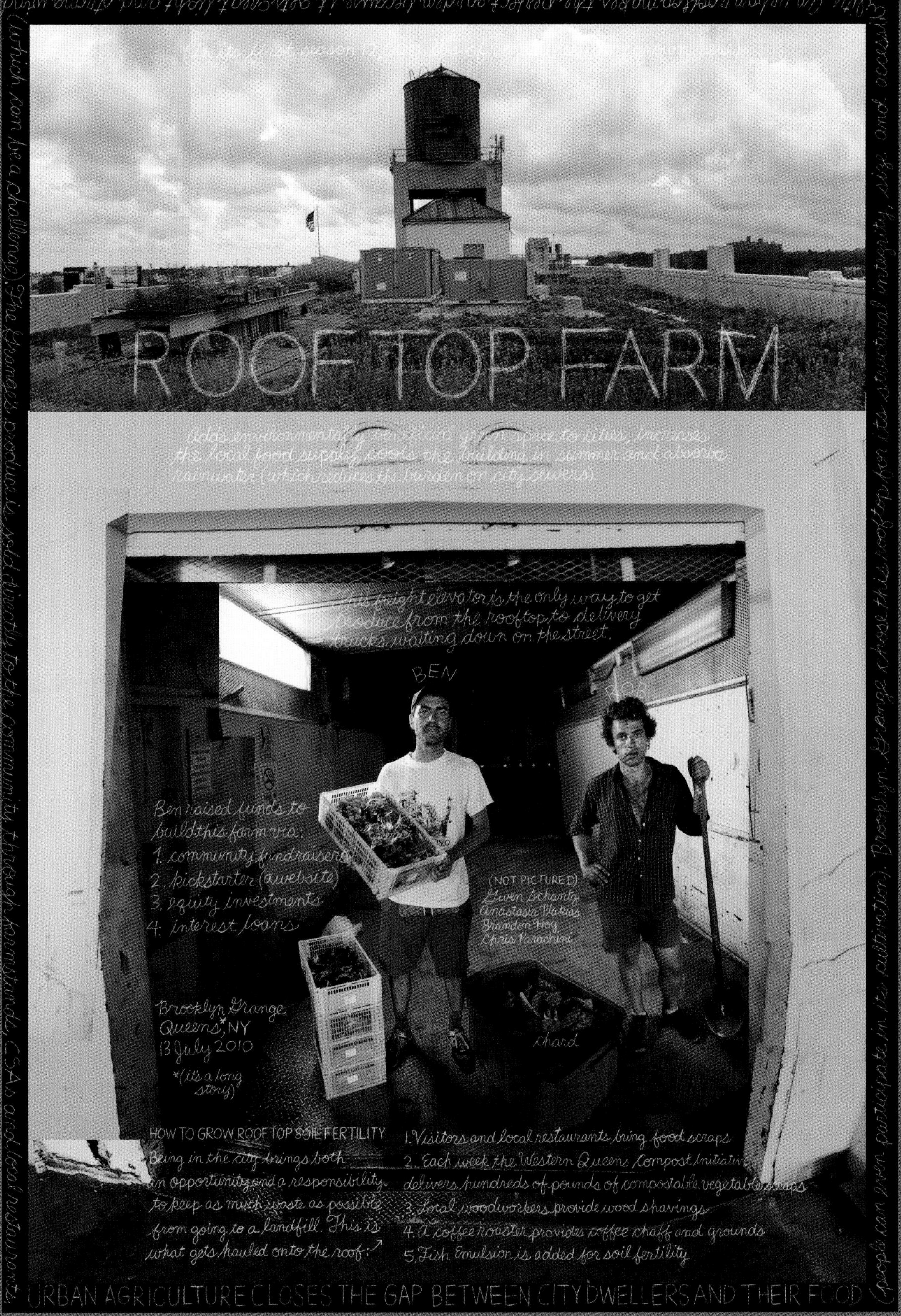

URBAN AGRICULTURE CLOSES THE GAP BETWEEN CITY DWELLERS AND THEIR FOOD
(people can even participate in its cultivation). Brooklyn Grange chose this rooftop for its structural integrity, size and accessibility.
An urban rooftop makes the perfect garden because it gets great light and strong wind
(which can be a challenge). The Grange's produce is sold directly to the community through farmstands, CSAs and local restaurants.
(In its first season 12,
ROOF TOP FARM
Adds environmentally beneficial green space to cities, increases the local food supply, cools the building in summer and absorbs rainwater (which reduces the burden on city sewers).
This freight elevator is the only way to get produce from the rooftop to delivery trucks waiting down on the street.
BEN
ROB
Ben raised funds to build this farm via:
1. community fundraisers
2. kickstarter (a website)
3. equity investments
4. interest loans
(NOT PICTURED)
Gwen Schantz
Anastasia Plakias
Brandon Hoy
Chris Parachini
Brooklyn Grange
Queens,* NY
13 July 2010
*(its a long story)
chard
HOW TO GROW ROOF TOP SOIL FERTILITY
Being in the city brings both an opportunity and a responsibility to keep as much waste as possible from going to a landfill. This is what gets hauled onto the roof: ↗
1. Visitors and local restaurants bring food scraps
2. Each week the Western Queens Compost Initiative delivers hundreds of pounds of compostable vegetable scraps
3. Local woodworkers provide wood shavings
4. A coffee roaster provides coffee chaff and grounds
5. Fish Emulsion is added for soil fertility

Handwritten annotations on photograph:

greasy pill. They might be serious cooks who have discovered that farmers food tastes better, or a concerned mother suspicious of a food system that recalls large quantities of vegetables each year. Regardless of the original intent, as people begin growing food, they embrace these ideas.

Regiment Spinach

French Breakfast Radishes

Flat-leaf Italian Parsley

Surrey Arugula

Sp—

T-shirt text: Colin grows food where he's not supposed to.

Hakurei Turnips

Roof of Bastille Café
Ballard Neighborhood
Seattle, WA
9 May 2011

WHERE COLIN PLANTS GARDENS
1. Single family homes
(either in front, side
or in backyard)
2. Restaurant rooftops
3. Restaurant patios
4. Condominium rooftops
5. Apartment complex
courtyards

YIMBY
Yes, in my Backyard (or on—
The community-based support of a new concept
their quality of life and connects them to their

COLIN MC CRATE BELIEVES IN THE PRINCIPLE OF "HYPER-LOCAL FOOD PRODUCTION", WHICH HE DEFINES AS, "FOOD GROWN, PRO—

ROOFTOP GARDENING CHALLENGES
1. Containers or beds have limited soil depth, making soil management difficult
2. Soil is less biologically active than in-ground gardens; less robust soil ecosystems make insect or disease problems harder to manage
3. Rooftops are much windier, requiring the construction of more rigid garden structures
4. Extreme rooftop temperatures require the use of shade structures to protect high value crops like salad greens
5. Crops require cold-weather protection to remain viable throughout the year

Nancy Butterhead Lettuce

Peppergrass

Deer Tongue Bibb Lettuce

Winter Density Romaine Lettuce

In Los Angeles, some areas were actually designed to encourage urban agriculture. In the 1920s, Charles Weeks, a land developer and devout agriculturalist, was courted by the Los Angeles Chamber of Commerce and given land to build the "Weeks Poultry Colony" in what is now Winnetka, a suburban community in the city's San Fernando Valley. His vision, an ambitious mix of utopian philosophy, spiritual naturalism, and practical small business tips for would-be entrepreneurs—all perfectly expressed in his twin manifestos *Egg Farming in California* and *One Acre and Independence*—promised huge financial rewards for urban farming on small land parcels. Weeks was almost right. The poultry industry prospered for ten years, only to go bust during the Great Depression, with many of his oblong land plots later converted to suburban use.

Enter Gary Jackemuk and Craig Ruggless seventy-five years later. They buy a house on a sleepy Winnetka street, formerly part of the Weeks Poultry Colony, tear out their backyard, and build a **SUBURBAN MICROFARM** complete with chicken coops and greenhouses. They now give cooking classes to local residents, sell heirloom seeds from racks in their living room, and grow produce for a local restaurant appropriately named Forage.

Together they're partially upholding Weeks's prophecy that one day urban agriculture would reign in the San Fernando Valley.

HYPERLOCAL FOOD PRODUCTION is "food grown, processed, and consumed at the neighborhood level of a community."
—Colin McRate, Seattle Urban Farm Company

DELLARAGIONE MINESTRA
Recipe by Craig Ruggless
Serves 4

My great-grandfather, Giuseppi Dellaragione, would say, "Minestra. It makes a weak man strong." Minestra is a stewlike soup, made rich with beef short ribs and "thick" with big handfuls of spigariello greens. Spigariello, more properly called *Cavolo Broccolo a Getti di Napoli*, is an heirloom leafy green that tastes like broccoli. But unlike broccoli, you eat the leaves instead of the flowers. Braise the beef the day before you plan to serve the soup, so it has time to develop a rich flavor. I know some people will think it is crazy to take two days to make a soup, but taking the extra time is worthwhile. A layer of fat will form after refrigeration; remove before reheating the soup before serving.

INGREDIENTS:
6 tablespoons olive oil plus extra for drizzling
3 to 4 pounds beef short ribs
Salt
Freshly ground pepper
1 to 2 large onions, roughly chopped
2 carrots, cut into 1-inch pieces
2 stalks celery, cut into 1-inch pieces
2 bay leaves
Whole peppercorns
About 8 cups beef stock or water
Several bunches spigariello greens, thinly sliced crosswise; kale, turnip, or a combination of mixed greens can be substituted
Pecorino or Parmigiano Reggiano for serving

DIRECTIONS:
1. Heat 6 tablespoons olive oil in a Dutch oven or large heavy-bottomed pot with lid over medium-high heat. Season the short ribs with salt and pepper. Once the oil is hot, place the short ribs in the pot and brown all over, in batches if necessary to avoid crowding the pan.

2. Add the onions, carrots, and celery to the browned short ribs, and saute until the vegetables are translucent. Add two bay leaves and a few peppercorns.

3. Add enough stock to cover the meat and vegetables. Increase the heat to high and bring to a boil, then reduce to the lowest setting and simmer until the meat is tender, about 2 hours.

4. When the meat pulls apart easily with a fork, remove the pot from the heat, let it sit at room temperature until cool, then refrigerate, covered, for at least 4 hours.

5. Remove the layer of fat from the top of the soup with a spoon; discard. Place the pot over medium-high heat and return to a full boil, then reduce to a medium simmer and add the spigariello. Cook the greens until tender, 7 to 8 minutes, and adjust the seasoning. To serve, ladle a short rib and some soup with greens into a large bowl, drizzle with olive oil, and grate some cheese over the top. Enjoy with some good rustic bread and a glass of red wine.

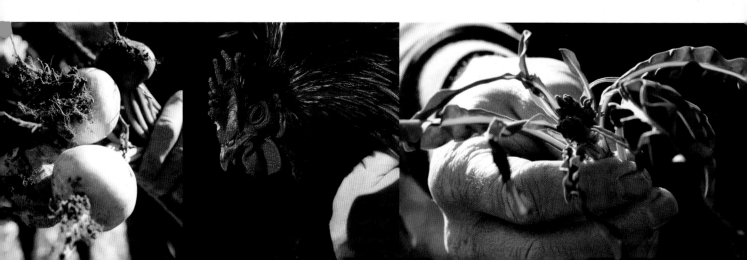

'Local is my own backyard.'

Winnetka Farm
Winnetka, CA
2 February 2011

→ hen house

→ Gary

a Dutch
Bornevelder
rooster →

"rapa da
orto bianco
Lodigiana"
white turnip
from Lodi, Italy

Craig
"barbola da
orto fondo
di Chioggia" →

"broccoli
spigarello"
leaf broccoli
southern Italy

mesclun mix →
"di mesclun"
mixed salad

(SUB)URBAN
MICRO FARM

a small farm in a mostly
residential community

CRAIG AND GARY STARTED THIS FARM IN THEIR BACKYARD TO GROW WHAT THEY COULDN'T FIND IN THEIR LOCAL SUPERMARKET.

Gary says, "In the last twenty years we've lost hundreds of varieties of vegetables from our tables because commercial growers simply stopped growing them." "The U.S. has basically one zucchini and one yellow squash now," adds Craig. "People of this generation don't know what they're missing." Craig, a young rock scout, grows nationally and they've both moved that is grown nationally and they've both moved.

PART IV
LIVESTOCK

The second half of the twentieth century has seen the consolidation of most aspects of our food system. Cattle, for example, are finished in industrial-sized Concentrated Animal Feeding Operations (CAFOs), which house thousands of animals.

I've spent my time on cattle ranches and in slaughterhouses, and I've even helped butcher animals, but I have no pictures of CAFOs. In many states, taking one would actually be illegal. They've passed "ag gag laws" and "veggie libel laws" to prevent you from seeing how your food is produced. Fortunately, this book is about sustainable solutions to the myriad problems created by industrial agriculture. CAFOs are not sustainable. They're the problem, not the solution. Most of the meat you eat, whether it's beef or pork or chicken, comes from a CAFO. Whatever you imagine CAFOs to be like, they're actually worse.

Creating a more sustainable alternative begins by understanding how these animals should live: what they should eat, and how they should be treated, slaughtered, and ultimately transformed into the food we eat. It's about grass-fed, pasture-raised animals that live outdoors and are treated humanely, without the use of antibiotics. It's a complex puzzle, one that requires returning forgotten pieces to local food systems, and reestablishing transparency and trust.

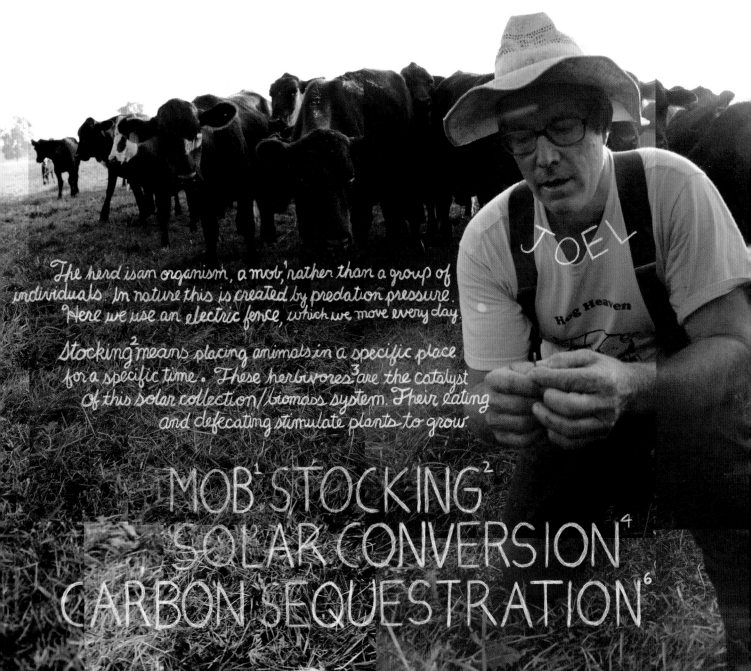

The herd is an organism, a 'mob,' rather than a group of individuals. In nature this is created by predation pressure. Here we use an electric fence, which we move every day.

Stocking[2] means placing animals in a specific place for a specific time. These herbivores[3] are the catalyst of this solar collection/biomass system. Their eating and defecating stimulate plants to grow.

MOB[1] STOCKING[2]
SOLAR CONVERSION[4]
CARBON SEQUESTRATION[6]

Lignin is the glue that holds plant cellulose together; as a plant matures green manuring. Nature does not let biomass drop to the soil surface until it burns – its energy – and drives the soil food web.

Plants create bilateral symmetry at the soil horizon. When grazed they voluntar "pulsing" occurs exponentially as plants achieve their juvenile growth spurt. This re

This routine dumping of organic matter into the soil feeds the soil biota (e

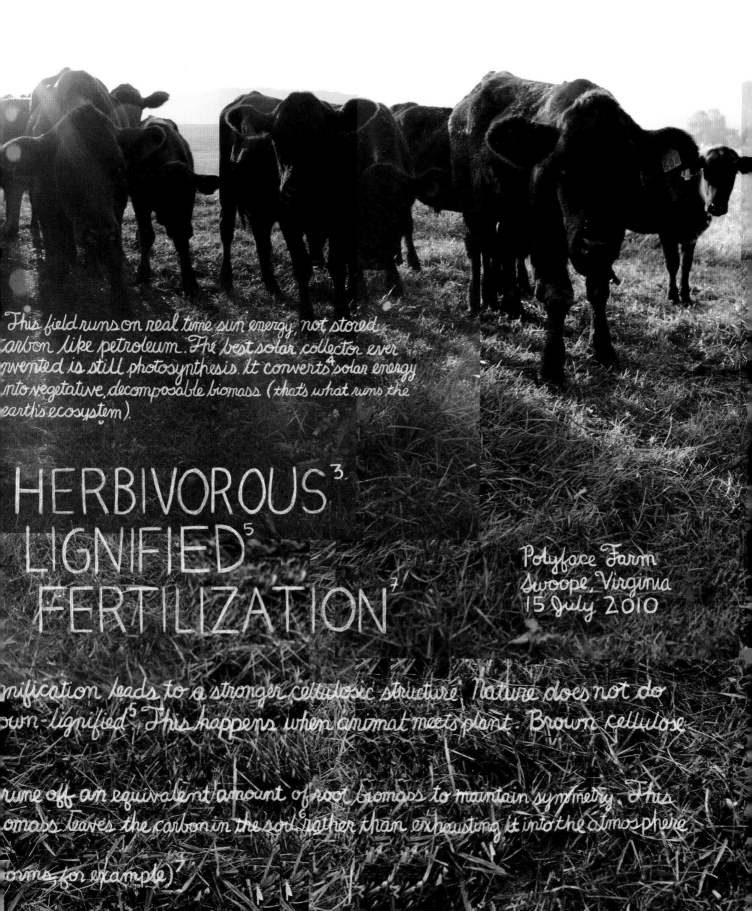

This field runs on real time sun energy, not stored carbon like petroleum. The best solar collector ever invented is still photosynthesis. It converts[4] solar energy into vegetative, decomposable biomass (that's what runs the earth's ecosystem).

HERBIVOROUS[3]
LIGNIFIED[5]
FERTILIZATION[7]

Polyface Farm
Swoope, Virginia
15 July 2010

...mification leads to a stronger cellulosic structure. Nature does not do ...own-lignified[5]. This happens when animal meets plant. Brown cellulose...

...rune off an equivalent amount of root biomass to maintain symmetry. This ...omass leaves the carbon in the soil rather than exhausting it into the atmosphere

...orms, for example)[7]

GRASS FARMING

Joel Salatin raises chickens and cows at Polyface Farm near Swoope, Virginia, but if you ask what he does, he'll tell you he's a **GRASS FARMER**.

Our day starts before sunrise. We walk in silence up a long, sloping field, passing a cluster of mildly curious cows as we approach two metal sheds. On closer inspection, I notice that the sheds have wheels, like on a school bus. Then come the chickens. Thousands of them. The chickens live on this pasture, and lay their eggs in these **EGG MOBILES** that Salatin moves each day.

Sustainable agriculture has no single figurehead—nor does this defiant, disparate movement have a center—but if it wants an able spokesperson, Salatin would be a safe bet. He's a professional contrarian, a knowledgeable agricultural apostate who not only practices what he preaches but has the rare capacity to explain it to others. While his summers are devoted to farming, his winters are spent literally barnstorming the country—from grange hall to farm to classroom—as he expounds on the joys of grass farming.

The science of it is simple enough. Grass is a solar collector. It uses photosynthesis to transform the sun's rays into chlorophyll. When cows eat grass, they convert this energy into protein and fat.

Field grass grows in three phases. The first, which Salatin refers to as the "diaper phase," is typified by slow development. This is followed by a massive growth spurt, what he calls a "virulent, vibrant teenage phase," as grass converts solar energy into chlorophyll. From there grass goes into senescence, or in Salatin's words, "the nursing home phase." If the grass can be kept in that highly productive middle stage, where it continually captures solar energy and turns that into biomass, it will produce in abundance, but how would you maintain this herbaceous fountain of youth in a perpetual state of production, or what Salatin calls the "**BIOMASS ACCUMULATION ACCELERATION PHASE**"? Salatin does it with steel rods and wire. He places them in a line that bisects the field, with each rod twenty feet apart. Then he attaches a length of wire from rod to rod. When he gets to the end, he wraps the wire around a battery cell that is continuously charged by a small solar panel. By flipping a switch he suddenly has a portable electric fence. When Salatin removes the fence closest to the cows, they cross into fresh new pasture and graze. Their manure is left behind to fertilize the soil, a process hastened by the aforementioned chickens. Salatin rolls their egg mobiles into pasture previously occupied by cows. By spreading the manure as they walk, these chickens accelerate the growth of new grass. This daily cycle, one of moving fence posts and opening up new pasture, follows a formula based on observation. The size of new pasture is determined by knowing how much the cows need to eat. With the proper allotments set, this **ROTATIONAL GRAZING** continues until Salatin reaches the end of the field. Then he simply brings the cows and egg mobiles back to the top of hill and the process repeats.

Cows are herbivores. Just like every other grazing herd in the world, these cows are followed through their grazing fields by birds. They scratch through dung and peck parasites off herbivores (like the egret on the rhino's nose).

The birds in this field are chickens. These omnivores are a biological pasture sanitizer. As the electric fence moves down the field, placing the cows on fresh grazing land, the chickens follow behind, preparing the paddock for the next grazing cycle.

cows

electric fence

chickens

Hog Heaven

Polyface Farms
Swoope, Virginia
15 July 2010

GRASS FARMER

JOEL SALATIN'S "POLYFACE BOUQUET": 1. fall panicum; 2. pigweed; 3. chickory; 4. timothy; 5. fescue; 6. redtop; 7. orchard grass; 8. more redtop; 9. more pigweed; 10. narrow leaf plantain; 11. wide leaf plantain; 12. more orchard grass; 13. red clover. Everything his food & hands grows plentifully in this field (he picked this entire bouquet from by his feet)

"It's essentially a biomass restart button," Salatin explains, "operated with the choreography of a ballet, right here in the pasture with herding, moving animals. The cows move themselves. They feed themselves. They fertilize the ground behind themselves. **Every time you have to harvest with petroleum and machinery, there's a tremendous amount of labor.** But what I'm talking about—the entire infrastructure—consists of some electric fence that you put in a wheelbarrow and some plastic pipe to run water. That's the entire deal. I mean, you don't even need a machine. It's all run on real-time solar energy and solar dollars."

Fence lines are revealing. I suddenly notice Salatin's neighbor a few hundred yards away. His side of the fence is barren, the grass grazed down to its roots. Cows stand impatiently, waiting as he struggles to tip seventy-five-pound hay bales off his truck bed. That hay was grown, harvested, baled, then trucked to his farm. That's a lot of energy. Watching a farmer go to such expense trucking feed in for animals that stand on perfectly fertile pasture—after what I've been doing the past hour—well, it just seems comical.

"Are you guys friends?" I ask.

"Our neighbors think we're bioterrorists," Salatin observes. "Why do so few people embrace the truth? I mean, that's the question of the ages. Why? Is it because grandpa did it that way? Because the [USDA] doesn't promote it? Because the research at the USDA is financed by large corporations that make their income by making sure farmers spend a lot of money?" Salatin wonders. "The agriculture press, the agriculture research, the agriculture media—and shall we say, 'agricultural subculture'—is literally immersed in an anti-ecology mind-set. It just takes a lot of personal savvy to walk away from all that. If people really began to embrace what we do, it would completely invert the profit, power, position, and prestige of the entire food and farming system. That is a tremendous amount of inertia to flip over."

Good ideas find fertile soil. They grow. After Google funds a number of information artworks for the Lexicon of Sustainability project, they ask me to design a restaurant at their Mountain View, California, headquarters. We etch terms from the project into glass windows. We turn our images into wallpaper that cover every available surface. We install video monitors to show our films. We then consult with their chefs to create a local menu. One of them, J. C. Balek, a heavily tattooed San Francisco chef (see p. 30), pulls me aside after the restaurant opens, points at an image, and says, "What is this one about?"

The image shows farmer David Evans explaining **PASTURE MANAGEMENT**, a principle he learned directly from Joel Salatin. Balek ends up spending a week on Evans's Marin Sun Farm in Inverness, California. By gathering eggs, moving chicken houses, and rotating cattle, Balek learns the meaning of not only a new term but an entire farming philosophy. "My entire career was spent in high-profile kitchens," Balek tells me afterward. "But seeing the farm changed me as a person as well as a chef." When he goes back to Google—and even later when he becomes head chef at Palantir, a fast-rising Silicon Valley tech company—he decides to serve food only if he can verify the producers' pasture management strategy. It does take great energy, as Salatin says, to overcome inertia and "flip" an entire food system over. But it can happen, one person at a time.

Marin Sun Farms
27 October 2010
Inverness, CA

PASTURE MANAGEM

To maintain the pasture as a viable renewable resource David u
poultry as tools to harvest the solar energy stored in these gr
it into nutrient-dense food. This process also maintains past

ruing beneath them allow the houses to be moved each morning providing the chickens fresh pasture.

erations. While his parents primarily raise cattle, David has int

He's a pasture farmer whose family has worked these headlands for four gen

...ENT.
...ttle and
...and convert
...iodiversity.

...ID IS THAT THEY THINK HE'S A POULTRY FARMER. (HE'S NOT.)

Higher in Omega 3s

protein

lower in saturated fat

higher in Vitamin E

grass

GRASSFED BEEF

ACIDOSIS

BLOAT ULCERS

DIARRHEA

GRASS FED

It once took four years to raise beef cattle. Now it's down to sixteen months. Cattle are raised on grass, then finished on corn—plus a mix of antibiotics, hormones, and protein supplements—on industrial-sized feedlots. Supplemental grain has always been part of a domesticated cow's diet, but cattle feed primarily based on corn is a relatively new phenomenon.

The main difference between **GRASS-FED** and **CORN-FED BEEF** has to do with physiology, or how a cow's gut works. Cattle are ruminants. Their stomachs have multiple chambers. With grass-fed cattle, each chamber plays a role in allowing cows to digest the cellulose in grass. Grass-fed beef is lower in saturated fats, and higher in omega 3s and vitamin E.

The delicate chemistry in a cow's rumen was never designed for a corn-based diet. It creates too much acidity and leads these animals to suffer from acidosis, bloat, ulcers, diarrhea, and a weakened immune system.

Cattle are grazing range dwellers. They eat grass. Rangelands cover nearly 40 percent of the United States and are the single largest ecotype on earth. The amount of available forage on these lands depends on a variety of factors, including climate and geography, but proper rotation grazing practices actually help these rangelands regrow by triggering photosynthesis, which pulls carbon dioxide out of the atmosphere and captures carbon in the soil. **CARBON SEQUESTRATION** is a major weapon in the battle against climate change. That's why some consumers buy grass-fed beef; it makes for healthier cattle and may even be better for the environment.

After unsuccessfully battling the industrial beef industry for twenty-five years growing conventionally raised beef for the commodity market in St. Francis, Kansas, Mike Callicrate set off in the opposite

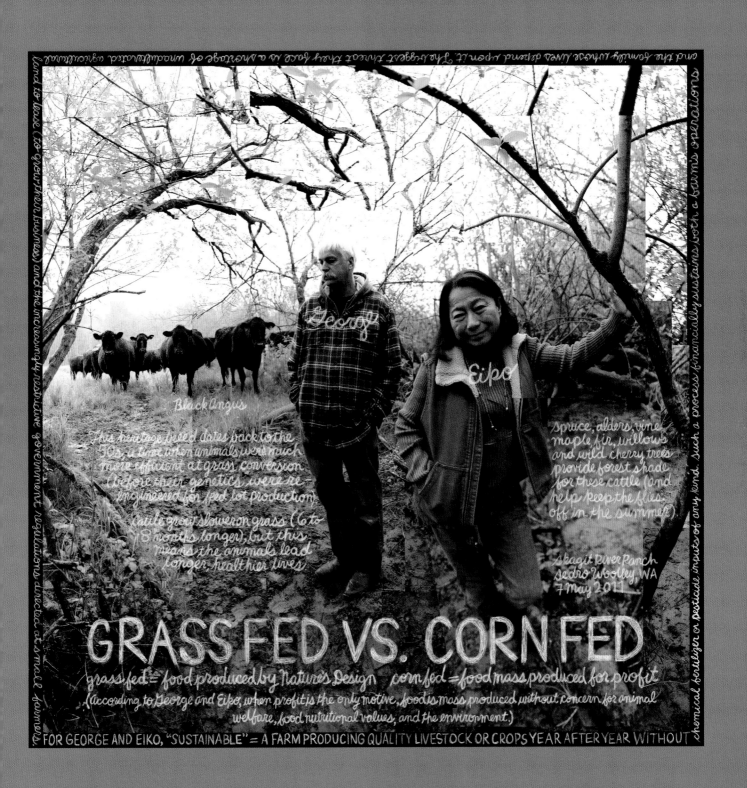

Black Angus

George

Eiko

This heritage breed dates back to the 50s, a time when animals were much more efficient at grass conversion (before their genetics were re-engineered for feed lot production).

Cattle grow slower on grass (6 to 8 months longer), but this means the animals lead longer, healthier lives.

spruce, alders, vine maple fir, willows and wild cherry trees provide forest shade for these cattle (and help keep the flies off in the summer).

Skagit River Ranch
Sedro Woolley, WA
7 May 2011

GRASS FED VS. CORN FED

grass fed = food produced by Nature's Design   corn fed = food mass produced for profit

(According to George and Eiko, when profit is the only motive, food is mass produced without concern for animal welfare, food nutritional values, and the environment.)

FOR GEORGE AND EIKO, "SUSTAINABLE" = A FARM PRODUCING QUALITY LIVESTOCK OR CROPS YEAR AFTER YEAR WITHOUT chemical fertilizer or pesticide inputs of any kind. Such a process financially sustains (with a farm's operations and the family whose lives depend upon it. The biggest threat they face is a shortage of unadulterated agricultural land to lease (to grow their business) and the increasingly restrictive government regulations directed at small farmers

managed holistically, balancing the needs of people, animals, and plants, while remaining ever mindful of his responsibility to both his customers and the community.

Los Angeles

Pacific Ocean

2012
Dodge Ram 2500 4X4

JOE
(son)

		PICK UP TRUCK		CULL COW PRICE (APPROX.)	
1985		$3,800	÷	$480	=
2012		$51,000	÷	$840	=

"COW TO PICKUP TRU

The number of cull cows required to be
of an integral piece of ranch equipment

HOW CAN JOE MORRIS MAKE A PROFIT WHEN THE PRICE OF EVERYTHING GOE

Morris Grassfed Beef
Sandhill Field
Circle P Ranch
Watsonville, CA
30 December 2013

San Francisco →

1985
Dodge Ram 2500 4X4

RICHARD
(father)

CK' INDEX

to cover the price
... a pickup truck

... EXCEPT FOR THE PRICE OF A COW?

Joe maintains a herd of Angus, Red Angus and Hereford grassfed cattle on coastal rangeland just outside Watsonville, CA

direction. "We legislated. We litigated. Everything failed," he recalls. "The only thing I could think that might work was the alternative food system; producing healthy food from a sustainable production model."

Callicrate rebuilt his business, starting with his own regional distribution system, bypassing middlemen and the entire beef industry's centralized processing and distribution apparatus by selling direct to consumers.

His neighbors weren't so lucky.

"The problem with these guys is that they believed in the industrial model," Callicrate observes. "They were told by universities and big corporations that they had to get big or get out. Now, they've got too much debt and too many animals. These aren't real cattle producers who believe in **STEWARDSHIP** and **HUSBANDRY**. They're businessmen feeding cattle, margin operators who bought into the industrial model because their paycheck depends on it, but, boy, are they having less and less fun each year, as these cattle continue to lose money and their equity disappears. They know they're on their way to **CHICKENIZATION**."

Joe Morris, a cattle rancher in San Juan Bautista, California, was born wearing cowboy boots, or so his mother claims. He daydreamed about becoming a real California vaquero, but when he got older, local ranchers tried to talk him out of the trade. They knew the hard facts. **The costs of doing business kept going up while the revenue from managing livestock didn't. Call it the COW TO PICKUP TRUCK INDEX.**

"It was simple economic calculus," Morris explains. "In 1970, you could sell cull cattle—cows that no longer have offspring—for $500 on the open market. A pickup at that time cost $3,500 to $3,000. Today, you can sell that same cattle for a little bit more—basically about $800—and a pickup now costs $50,000."

PASTURE-RAISED

We have chickens on our farm. Exotic Lakenvelders and Barnevelders. Easter egg–laying Auracanas. Old-time classics like Wyandottes, Buff Orpingtons, and Rhode Island Reds. When friends visit with their kids, the children often make a dash for the coop. Our young daughter, guided by only the vaguest notion about 4-H, took it upon herself to train the chickens. They now stop and sit when she approaches, waiting obediently to be picked up and carried about the property. When I hand the parents freshly laid eggs, they recoil slightly. They buy eggs. They eat eggs. But they don't necessarily visualize where those eggs come from.

We are largely disconnected from the food we eat. Despite the pastoral narratives our egg cartons depict, the real story behind our eggs is probably much different. When we get past the pictures, we find words. I remember when the term **CAGE-FREE** first appeared. I had no idea what it meant, but figured the egg would probably taste better and the chicken might have a better life as well. The term triggered another unexpected result. For the first time, people wondered about the living conditions of "caged" chickens. It turns out they were pretty dismal. These chickens were crammed into battery cages, packed so tightly together their bodies could scarcely turn, with lives so stressful their beaks were snipped to prevent them from pecking one another to death. **When consumers voted in favor of cage-free eggs, the industry was forced to change, but how much?**

That's what I ask Alexis Koefoed of Soul Food Farm in Vacaville, California, and she says, "I remember the first time I heard the term *cage-free* was when I read it on a carton in a grocery store. I got kind of excited. I thought maybe there were free chickens running around. It took a little research, and I found out it didn't mean much. Actually, the chickens were still in big huge houses on industrial factory farms. I didn't think it was anything more than just good marketing, so we think things are nicer and sweeter and kinder than they really are."

My friend David Evans agrees. His pasture-management strategy calls for the continued rotation of chickens and cattle on Northern California headlands I've known

since childhood. As a fourth-generation farmer he's seen, more than most people, how the poultry industry has changed. "A definition for *cage-free* is a simple one," he says. "It's 'not raised in a cage.' That's really all it means. It doesn't say anything about the environment they *are* in, just one they *aren't*. Clearly, we would hope that the environment is better for the chicken, but if it is standing in its own muck and still can't turn around, it begs the idea, where do we go from there?"

Consumers caught on. They recognized that cage-free eggs only offered a marginal improvement in a chicken's quality of life. Then the term **FREE-RANGE** appeared. "*Free-range* was a term I really loved," Koefoed recalls. "I really thought it meant something for quite a long time. Finally, I learned that that term had been hijacked as well."

"'Freedom' is a very American ideal," Evans observes. "When we apply that to a chicken it's like WOW! This chicken is 'free to range.' Well, it's free to range in a certain-sized space that isn't very big. It's much smaller than a chicken would roam in its natural environment."

After weighing her options, Koefoed decided that while *free-range* was confusing, it was still meaningful to put on her egg cartons. Her customers revolted. "People were just so disturbed by the term that we took it off," Koefoed remembers. "We realized we couldn't fight against the marketing giants who were using *free-range* as a term to sell more eggs even though they were the same old industrial chicken companies with confined animals."

Then she discovered the term **PASTURE-RAISED**. "I just love that term," Koefoed says, "because when you say 'pastured' you immediately think of a field so it really explains, very clearly, in one word, that the animals are outside, meaning 'grass,' meaning 'bugs,' meaning 'sunshine.' So I think it's a really good term to define what someone is doing, whether it's chickens or any other animal."

David Evans agrees, but thinks we still aren't done. "So we go 'cage-free,' we go 'free-range,' and now 'pasture-raised,'" he says, "and soon we'll go beyond that.

Stone Barns Center for
Food and Agriculture
Pocantico Hills, NY
26 June 2012

egg mobile
(moved once
a day)

pasture

poultry

CRAIG

MEGAN

PASTURED POULTRY

The raising of pastured poultry embraces "humane, people-friendly, environmentally enhancing" practices. After feathering out (and as the season permits), the poultry are given constant access to fresh growing palatable vegetation, with movable or stationary houses utilized for shelter.

THE EGGMOBILE PROVIDES:

1. shelter from the elements
2. protection from predators
3. a clean, private space to lay eggs
4. a place to roost at night

WHY SHOULD PEOPLE BE WILLING TO PAY MORE FOR PASTURED POULTRY? CRAIG HANEY SAYS, "DOING THINGS RIGHT TAKES MORE TIME.

We'll keep trying to define these terms so customers can support the type of food production models they want."

Movements promoting "good food" succeed when the messengers become their messages, when their foremost practitioners embody the language of sustainability, and when a farmer doesn't just farm. "We're small farmers in a new world," Koefoed concludes. "We don't just farm. We're educators and we're learning to be marketers so we can hold on to the authenticity of words and take them back from the big corporations." That's something to think about the next time you buy a dozen eggs.

HATCHERIES and FEED

When I photograph farms I'm often left to wander. After all, taking pictures is only interesting if you're the one with the camera. I once visited a well-known Northern California poultry farm. "You have to go there," I was told. "These guys are *amazing*. They have the *best* eggs and most *amazing* chickens in California. So I wander. I take pictures. Thousands of truly happy chickens follow me around. They are as interested in me as I am in them. So many chickens and yet I find no hatchery. So I ask a worker.

"Wednesday," he tells me.

Farmers are idiosyncratic. They often refer to their farm structures with the most peculiar descriptions, so I look for a building called "Wednesday." I search, but I don't find it, so I ask the worker again, "Where is this hatchery?"

"Wednesday," he repeats.

I just stare at him. There is no building called "Wednesday." Of this I know.

"It comes on Wednesday," he tells me firmly. "Wednesday morning with the FedEx."

And that's when I learn something pretty interesting. Sometimes even the most local of foods actually doesn't start out that way, and if it's chicken, it probably begins at a Midwest hatchery.

Petaluma, California, where I live, used to be the egg capital of America. Each year we still have a Butter and Eggs Day. There's a parade. Half the town marches. The other half watches and cheers. There's even a beauty pageant for babies dressed in chicken costumes. It may be exploitative, but it's also cute. I write these very words in a barn lined with planks taken from a fallen chicken coop. A few still stand outside my window. This was a poultry community. **HATCHERIES** defined our town. We were good at it. Nowadays, poultry farmers don't hatch out their own chicks. They come on Wednesday, in cardboard boxes FedEx delivers from the Midwest.

I could tell you the same story about what animals eat, chickens in particular. I visit another farm. Yet

create a healthy bird that produces a healthy egg—I don't think it's enough to be simply being certified organic.

George

"If people don't care eat, they obviously wo what animals eat."

WHEN GEORGE AND EIKO DISCOVERED THAT THE "ORGANIC"

186

So, I really don't trust their grain. There must be a reason Chinese grain is so much cheaper. I want to provide inexpensive Chinese grain gets inspected by USDA, and still half of them get rejected for contamination."

Skagit River Ranch
Sedro Wooley, WA
7 May 2011

EibO

ORGANIC FEED

A feed designed to mimic a chicken's natural diet.

I. trace minerals
II. seaweed
III. diverse grains

IV. no corn (especially Chinese raised corn)
V. rock powders

VI. carbon humates
VII. sprouted wheat, barley and oats

t they
are

George says, "Only 1-2% of Chinese grain gets inspected by USDA, and still half of them get rejected for contamination."

EN FEED THEY BOUGHT CONTAINED UNVERIFIED "ORGANIC" SOY MEAL FROM CHINA, THEY DECIDED TO MAKE THEIR OWN.

again, I'm left to wander. This time I come upon a barn with the door left open. Inside I discover bags of organic chicken feed from China.

FEED is the single greatest input cost in any poultry operation. The cheaper the feed, the higher the profit. A chicken's diet is based on corn and soy. The single largest—and cheapest—supply of organic soy comes from across the Pacific Ocean on tanker ships.

Can organic poultry feed mean less **FOOD MILES**? I find the answer on Utopia Road in Sedro-Woolley, Washington. George and Eiko Vojkovich raise pastured poultry on an outdoor diet rich in bugs, worms, and grass, lots of grass. To mimic what nature provides and augment this feed regimen, the Vojkoviches created a device that produces one hundred pounds of sprouted wheat, barley, and oats a day. To that they add trace minerals, seaweed, rock powders, and carbon in the form of humates. They also include diverse grains like spelt and amaranth. Two things you won't find? Chinese soy or corn.

How do these eggs taste? If you're in Seattle, look for the Vojkoviches at an area farmers' market. Get there early and prepare to wait in line fifteen minutes for a dozen eggs.

Whatever an island doesn't grow or produce, it ships in. Cattle production on the Hawaiian Islands presents an interesting conundrum. **What's cheaper? Shipping grain for cattle feed from the U.S. mainland to Hawaii or sending cows across the Pacific to a Texas feedlot, where they are finished on a grain diet, slaughtered, then sent back to Hawaii?**

If you guessed that it's cheaper to send feed, you're wrong. The answer, obviously, is to find local feed sources. When Bill Cowern converted a Kauai sugarcane plantation into a tree farm, he noticed that the dense guinea grass growing beneath his trees had unusually high protein levels. He developed a system to cut the grass and transform it into nutrient-rich **GRASS CUBES** that are now used as cattle feed by local ranchers.

(Angus and Brahman crosses)

"Hawaiian Mahogany Farm
Lawai (Island of Kauai) Hawaii"
14 Dec 2012

SHADE-GROWN GRASS
HAS MORE PROTEIN
Lower light levels beneath the forest
canopy cause shade grown guinea grass
to produce more phytoplasts. More
phytoplasts mean more protein than that
produced by open grown guinea grass.
(15% vs. 8%)

Bill converted a 100
year-old sugarcane plantation
into a grove of
Albizia and rainbow gums

GUINEA GRASS
Panicum maximum

(then discovered the
grass growing beneath it)

GRASS CUBES

wild guinea grass that is harvested, chopped to 1½ lengths, dried to 15% moisture,
then fed through a cuber. It can be used as local feed replacement for cows and horses

90% OF HAWAIIAN CATTLE ARE SHIPPED TO THE MAINLAND AND FINISHED ON FEEDLOTS. BILL COWERN MAY HAVE ANOTHER SOLUTION.

8 Mc of grain to grow 1 Mc of beef. To ship 8 Mc of feed cost $1,40 a Mc. 1 Mc of beef grown from the same grain that fed that animal cost $1.20 per Mc. For beef to be local to Hawaii, it's critical that they develop local feed supplies, like grass cubes. Do the math: it's cheaper to ship a 400 Mc. cow to a Colorado feedlot than to ship 4 tons of feed to Hawaii. Why? It takes

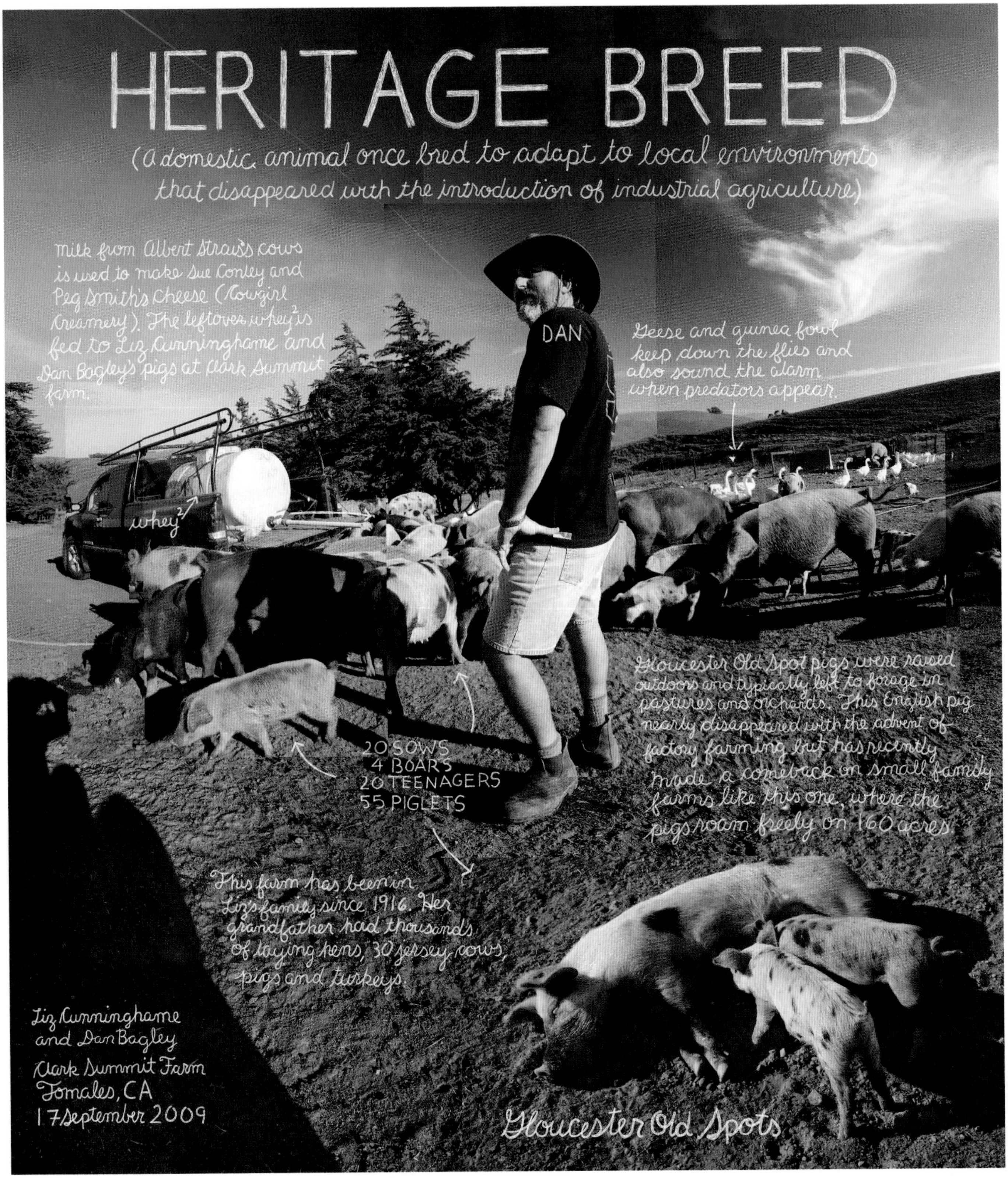

HERITAGE BREED

(a domestic animal once bred to adapt to local environments that disappeared with the introduction of industrial agriculture)

milk from Albert Strauss cows is used to make Sue Conley and Peg Smith's Cheese (Cowgirl Creamery). The leftover whey is fed to Liz Cunninghame and Dan Bagley's pigs at Clark Summit farm.

whey?

Geese and guinea fowl keep down the flies and also sound the alarm when predators appear.

DAN

Gloucester Old Spot pigs were raised outdoors and typically left to forage in pastures and orchards. This English pig nearly disappeared with the advent of factory farming but has recently made a comeback on small family farms like this one, where the pigs roam freely on 160 acres.

20 SOWS
4 BOARS
20 TEENAGERS
55 PIGLETS

This farm has been in Liz's family since 1916. Her grandfather had thousands of laying hens, 30 jersey cows, pigs and turkeys.

Liz Cunninghame
and Dan Bagley
Clark Summit Farm
Tomales, CA
17 September 2009

Gloucester Old Spots

HUMANE TREATMENT

Mark Newman lives in a small town called Myrtle, Missouri. It's on a map but the town has no stop signs or traffic lights. Just a few single-story buildings done up in the classic porch-and-awning style, clustered beside a two-lane road. It's a place with more churches than supermarkets. Newman raises **HERITAGE BREED** Berkshire pigs here. *Outdoors.* Not *indoors.* He did the latter for nearly twenty years in northwest Iowa, until he visited England, saw how pigs were raised, rediscovered the neglected principles of **STOCKMANSHIP**, and had his own "Road to Damascus" moment. But I'm getting ahead of myself. Mark Newman raises pigs outdoors.

"We raise about one hundred pigs per pen and our pen space is basically five acres," he explains. "In modern pork production, pigs are allowed twelve square feet, not twelve hundred square feet per pig. So our pigs have plenty of land and shelter. They have water. They have feeders in the fields. They have the ability to run and play and dig in the dirt. In the end it makes for a better-quality product."

Understanding what type of pork products to buy requires learning about what they eat and how they're raised. It also requires understanding how they aren't raised, which gets back to the indoor pork production model.

"I was very good at teaching people how to raise pigs indoors," Newman recounts. This was in the early eighties, when farmers started sow cooperatives across the country. "Ten local farmers would get together and house one thousand sows indoors, in one spot," he continues, "and basically raise the pigs up to feeder pig size, then take them back to their individual farms for finishing. As time went on, these operations just kept getting larger and larger. In the nineties we went from family farms that were doing two hundred forty sows to eight hundred sows. Pretty soon two thousand sow barns were popping up. Today, five thousand to ten thousand sows per farm is not unusual." In fact, Newman reckons that 99 percent of the pigs in this country are raised indoors. It's what you'd call an industrial food system.

During that period he was a vocal proponent of placing sows in **GESTATION CRATES**, which he claims was for humane reasons. "Whenever you put ten sows or fifteen sows into a small pen, you get a situation where you have two very dominant sows and two very weak sows. Those two on the bottom get picked on by the other ten in the pen."

"So you raise pigs in crates to protect them," I reply. "Still, it doesn't seem very humane."

"I kind of agreed with confinement at one point in my life. But then I said, you know, we're concentrating these animals so heavily and I mean, the disease situations in some of these buildings are enormous."

Then Newman went to England.

"You know, everybody has a spot in his inner self," he explains, "what they think is the right way to do something. I was driving across the Midlands of England, south of Bedford on the Avon, and I saw these huts

outside in a field." Newman stopped, met the farmer, and spent a few days studying how he raised pigs.

"I saw that the animals were much more content, much happier, and the **STOCKMANSHIP** was so much better. Today, in America, one limiting thing—whether it's in the cattle industry or pork industry—is the loss of stockmanship: People don't really know how to read and work with animals. We happen to be fortunate. We have a daughter. You know the term *horse whisperer?* We always called her the 'pig whisperer.' She could process pigs in a farrowing hut, with mama right in there with her, and never have a problem. I mean, some people have a way with animals that other people don't.

"Stockmanship is applicable in any situation," he continues. "It doesn't matter whether it's on our farm with pigs being raised outside or in a total indoor confinement operation. Animals have rights. **We—as farmer and producer—have the responsibility to raise these animals humanely and raise them the right way.** I don't think you want to buy a product where you don't feel the animal was raised the way it should be."

Newman's decision to share his farming practices with consumers led him to Adele Douglass. "The first time I saw a pregnant sow in a stall, I was absolutely stunned," she explains. "I loved 'Old McDonald Had a Farm' and I had this very naïve perception of how farm animals were raised. Then I saw these pregnant sows and my first reaction was, 'I can't believe it. In all these years I had no idea this happened.' I guess most consumers don't either. That was my epiphany. That's when I thought, 'Somebody has to do something about this, so why not you?'"

Douglass was forty-seven at the time. She cashed in her 401(k), received funding from the Humane Society, the ASPCA, and individual donors, and set up Humane Farm Animal Care. The nonprofit now oversees the Humane Certified program, a set of standards that protects the physiological and behavioral needs of livestock. It's a voluntary plan, one that gives producers third-party verified certification to authenticate their humane farming practices and provides consumers with a certification seal that allows them to make purchases

in line with their values—and it's all done without government involvement. By 2012 widespread consumer support had led to more than 75 million animals being raised according to Humane Certified standards.

Andrew Gunther and his wife started the world's first organic poultry hatchery. "When we started farming, it was purely for profit," he observes. "Then we saw the benefits of alternative production. The epiphanic moment came when we saw that, despite not putting on herbicides and pesticides, our pastures didn't fall apart. That despite not feeding every animal everything the drug company was trying to sell us, we didn't see a significant decrease in production. Then you look at the benefits. The birdsong. The cleaner skin. The cleaner water. You can do this."

"We know the physiological behaviors of herbivores and cattle," he continues. "They do better on pasture and open ranges. We also know that the omnivorous species—hogs and chickens—are range dwellers; in their natural lives they live at the edge of jungles. They root and forage for food."

Gunther's notion of raising livestock on pastures or ranges—or what Mark Newman simply calls "outdoors"—is central to another form of **CERTIFICATION**: Animal Welfare Approved, or AWA. It's based on three principles. First, animals have to live close to their natural environment. Second, AWA only certifies people who own and have control of their animals. This ensures greater transparency and accountability. Third, their program covers animals for their entire lives, from birth to death. It's a holistic approach.

Like Adele Douglass, Gunther remains optimistic about the power of labels to shift not only people's buying patterns, but their consciousness as well. **"People are now understanding the connection between where their food comes from and how it is produced,"** he observes, **"and the connection between their own health and the health of the planet.** Once they understand the source of their food and the impact that production has on them, they'll start to make conscious and continuing change."

FOREST RAISED

Fresh-ranch Farm
Salisbury, MD
8 July 2010

These animals forage and hunt in their own food in the same habitat they would prefer if they were wild. The forest also provides them with shelter from the summer sun and cold winter winds; deciduous trees are the perfect passive solar home for a big.

(Pigs are curious
+ social creatures)

tree bark

Durocs

TED

WHAT PIGS FORAGE FOR

shrubbery
wild grapes
virginia creeper
mice
persimmons
dewberry
blackberry
hickory nuts
hollyberries
beetles
seeds
honeysuckle
ants
clover
lizards
insects
eggs
roots
elderberry

RAISING PIGS IN THE WOODS IS NOT AN ENTIRELY NEW PRACTICE, BUT WHAT IS NEW ABOUT THESE PIGS IS THAT THEIR FOREST FORAGING AREA ROTATES.

(using electric fencing). Ted lets them hit the wild forage hard until it dwindles, then moves them on to fresh ground. This mimics the pattern of wild animals, who are always moving to a fresh food supply, allowing the forest food supply to regenerate. Ted also enhances the wild forage supply by using forestry practices of selective thinning and clearing of trees and shrubs. Ted's pigs also get a grain-based pig feed, but this is only a fraction of their diet.

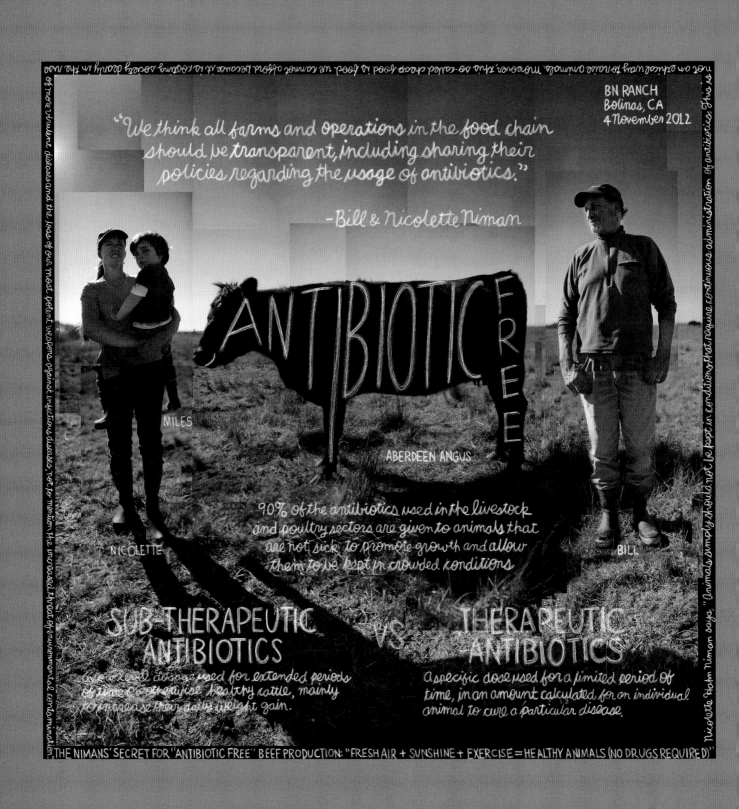

BN RANCH
Bolinas, CA
4 November 2012

"We think all farms and operations in the food chain should be transparent, including sharing their policies regarding the usage of antibiotics."

—Bill & Nicolette Niman

ANTIBIOTIC FREE

MILES

ABERDEEN ANGUS

NICOLETTE

BILL

90% of the antibiotics used in the livestock and poultry sectors are given to animals that are not sick, to promote growth and allow them to be kept in crowded conditions

SUB-THERAPEUTIC ANTIBIOTICS

vs

THERAPEUTIC ANTIBIOTICS

A low level dosage used for extended periods of time on otherwise healthy cattle, mainly to increase their daily weight gain.

A specific dose used for a limited period of time, in an amount calculated for an individual animal to cure a particular disease.

THE NIMANS' SECRET FOR "ANTIBIOTIC FREE" BEEF PRODUCTION: "FRESH AIR + SUNSHINE + EXERCISE = HEALTHY ANIMALS (NO DRUGS REQUIRED)"

ANTIBIOTIC-FREE

Nearly 80 percent of the antibiotics in this country aren't used on people. They're used on animals. Animals we eat. And most of these animals aren't even sick.

If the meat you eat isn't **GRASS-FED** or **PASTURE-RAISED**, if it isn't **CERTIFIED HUMANE**, if it didn't come from a local farmer or rancher you know from the **FARMERS' MARKET**, there's a good chance it was industrially raised on a **CAFO**. It's a centralized—and impersonal—model, one based on economies of scale.

As cattleman Mike Callicrate explains, "Industrial agriculture keeps animals in a stressful environment. When a corporation raising livestock wants maximum return on their investment, they pour more concrete. You want as many animals on that concrete as possible."

When an animal is sick, it's given **THERAPEUTIC ANTIBIOTICS**. It's like when you go to the doctor and get a prescription for antibiotics. You take them for a short period of time . . . then stop. But in CAFOs, antibiotics are often added to the daily feed regimen of perfectly healthy livestock.

"It's not natural for an animal to be in those tightly confined quarters," Callicrate continues. "As a result they're subject to disease due to stress and the environment they're in. And that disease is contagious, so they feed **SUBTHERAPEUTIC ANTIBIOTICS** to keep these animals from getting sick, but these also have the effect of increasing performance and weight gain, so it's a crutch for them."

That's right. **When animals are given antibiotics—even when they don't need them—they gain weight. And since a fat pig is worth more than a skinny pig, using antibiotics—even if an animal isn't sick—actually increases a farmer's profits.**

Except there's a problem. The misuse of antibiotics encourages the evolution of antibiotic-resistant bacteria. Bacteria that could be harmful to people. Plus, animals fed antibiotics create animal waste full of those very same antibiotics. This can contaminate our drinking water and ravage our ecosystems, which eventually allows these antiobiotics to work their way back to people . . . people like you.

HUMANE SLAUGHTER

Industrial agriculture is a work in progress. The proof can be found walking the halls of Colorado State University's Department of Animal Sciences, where giant posters announce the school's most recent research: "Characterization of *Escherichia Coli* 0157:H7 Shedding and Persistence in Feedlot Cattle"; "Effect of Feeding Frequency on Feedlot Steer Performance"; and "Use of blood lactate to measure swine handling stress from farm to processing plant: Relationship to pork quality," which features the contribution of Temple Grandin, a CSU professor who actually sees things from an animal's point of view. When she visits a feedlot, she grasps the architectural flaws that might spook an animal—abrupt pools of darkness, confusing patterns on walls or floors, building materials that make walking difficult—and proposes changes when necessary. This unusual gift, coupled with a doctoral degree in animal science, has made her one of the world's most renowned designers of animal containment systems. Nearly 70 percent of all cattle in the United States now pass through systems of Grandin's design.

We sit in a somber wood-paneled conference room lined with framed portraits of past department directors. This assemblage of august personages—dressed in their cowboy hats and western finery—looks down silently as Grandin recounts a story about an auditing program she designed to overhaul U.S. slaughter plants. It sat on a shelf for years, but she was patient. One day, McDonald's came knocking. This was in the 1990s, after the "McLibel Trial" saw Helen Steel and David Morris taken before a British judge for libeling the fast-food chain with claims critical of their business practices. McDonald's initially won the legal case, but lost the publicity war, so the company approached Temple Grandin.

"Heat softens steel; then people like me can run in and bend it, build it into pretty grillwork," Grandin explains, "because that's what you want to do: bend it into pretty grillwork, not make a mess. I was concentrating on trying to improve slaughter plants, so when McDonald's got some pressure on them and they had to do something on animal welfare, my ob was to take their executives through slaughterhouses. It was interesting, watching executives' eyes open. Watching them go from abstraction to something real. They were like 'Oh, man, there really is some bad stuff here.'"

"Did it surprise you that a company whose core business is selling cheap hamburgers had no idea how the meat was produced?"

"People don't know where anything comes from. The other thing I've learned about big corporations is the bigger they get, the dumber they get. They just totally lose control and lose contact with what's going on. That's true for every big corporation. I call it 'bureaucratic hardening of the arteries.'"

"When we started the McDonald's audit," Grandin notes, "the first thing we had to do was get people under control. A lot of problems were simple stuff like lack of repairs to the stunning equipment. But certain people like torturing and you just have to remove them. They shouldn't be handling animals. **Let's say we took away all the traffic cops. Can you imagine what a mess you'd have, with traffic and accidents and everything else? Well that's how animal handling is. It has to be constantly monitored.**"

Grandin sees "constant monitoring" as a way for the industry to police itself, but after telling Ag businesses to "open the door electronically" for years without success, Grandin simply started to make videos herself. "Video Tour of Beef Plant featuring Temple Grandin" and "Video Tour of Pork Plant featuring Temple Grandin" both offer unsentimental, straightforward explanations for how slaughterhouses work.

"I don't say, 'It's wonderful' or 'It's bad,'" Temple points out. "I just say, 'This is how it works.'"

The first step to fixing a food system is understanding not only how a food system works but that everything you eat tells a story, including some you might prefer not to know about.

SALMON
(primarily from tributaries of Frasier River)
sockeye - end of July
Pink salmon - mid August
Chum or Dog salmon - mid September
silver Coho salmon - end of september - October
Chinook - throughout the season
(Local trollers bring Neah Bay Chinook
from May 1 through July)

SHELLFISH
clams + mussels + oysters
available locally most months

Riley
the Reefnetter

SEA CUCUMBERS
September to May

HARO STRAIGHT HALIBUT
Indian tribes only: March to mid April

SPOT PRAWNS
Indian tribes: April to May
"Non-treaty" fishermen: June 2 0 to Augu-

DUNGENESS CRAB
Indian Tribes: throughout summer
"Non-treaty" fishermen: October to April

Lummi Island
6 May 2011

FISHING IN SEASON
a year in the (marine) life of the salish sea

PART V
FISHERIES

"We are disconnected from nature because
     we don't live in a way where the connections
     are apparent.

Most of us get food our entire lives from across a counter.

We don't know where it's been.

We don't know where it comes from.

It arrives like a letter that's been delivered
     to the wrong address.

We don't know where our drinking water originates.

We don't know anything about where our gasoline comes from.

We just don't see these things.

The connections are so disrupted from view that even if
     we wanted to know, it would be very hard
     and in some cases completely impossible
     to understand."

—Carl Safina, marine ecologist and author of *Song of the Blue Ocean*

TRACEABILITY

It doesn't matter where you are, or in what season. It could be snowing outside. You could be suffering through a heat wave. It makes no difference. You will still find tomatoes. The same can be said for chicken or steak, sausages or bananas. They'll always be there. But not fish. You might find fresh halibut or snapper or cod, or you might not.

"A lot of consumers don't realize that seafood is the last protein we hunt on a commercial scale," points out Jennifer Kemmerly, director of Seafood Watch at California's Monterey Bay Aquarium. "Here in the U.S., we don't hunt for beef anymore to feed the masses. We don't hunt for chicken, either, but we do hunt fish."

Carl Safina, a noted author and marine ecologist, agrees. "Fish are wildlife. They're not commodities; they're not belts or shoes. They don't just come from a warehouse," he says. "We often treat the ocean as a place where fish swim around until we catch them and they become useful, but they're always useful. There are birds that rely on fish. There are mammals in the ocean that rely on fish. It's all one big interdependent system. And we're part of that, too."

Consumers expect to find the fish they want whenever they want it. Entire industries—from restaurants to supermarkets—depend on a steady supply of fish protein. Barton Seaver, an internationally renowned chef turned sustainable food advocate, thinks these expectations are unrealistic.

"There's a reason why it's called fishing and not catching," Seaver explains. **"We don't always know what's going to come up."**

Cod once symbolized the wealth and abundance of Atlantic fisheries. Now it's an object lesson, one Seaver believes resulted from consumers having unrealistic expectations. "We want cod all the time," he says. "We have made cod king. In the Massachusetts statehouse, the cod hangs over the door. We just expect cod to be there. When we're very young we learn about object permanence. It's part of our psychological development. Even when an item is hidden we know it still

exists. Yet in our oceans we haven't learned the idea of object impermanence. **We hear all these stories about decreasing quotas, failing stocks, and failing fisheries, yet we still expect cod to be at the fish counter.** We don't understand that our actions are the principal drivers in making cod go away. Food is not an expectation. It's a hope, a desire, and the cause of much of our enterprise. It's important that we understand that. Nothing is guaranteed."

Just as cattlemen use rotational grazing to sequester carbon and manage rangelands, or farmers use a combination of compost, soil amendments, cover crops, and crop rotation to build topsoil and enhance soil fertility, fisheries require stewardship to manage the inordinate pressure placed on depleted fish stocks by unreasonable consumer expectations.

In 1968, Dr. Garrett Hardin introduced a principle called the **TRAGEDY OF THE COMMONS**. He claimed that if a community shares a common resource like a pasture owned by one and all, and if each farmer, motivated by enlightened self-interest, grazes his animals as much as possible to get the maximum benefit from this shared resource, other farmers will do the same. They will overgraze the commons until no grass is left.

The tragedy of the commons is a cautionary tale, one easily applied to the state of our global fisheries. Without safeguarding our oceans, certain fish, including **KEYSTONE SPECIES**, will be **OVERFISHED** or

> "**OVERFISHING** happens when fish are taken out of the water faster than they can reproduce to maintain their optimum population size. When fish stocks can no longer sustain a commercial fishery, fishermen have to find other jobs. The repercussions and the trickle-down effect—to both the marine food web and fishing communities—can be critical and severe."
> —Jennifer Kemmerly

EATING DOWN THE FOOD CHAIN[1]

Monterey Fish Market
Pier 33
San Francisco CA
9 December 2010

ALEJANDRO *the descaler*

RUSS[5] *the chef*

PAUL JOHNSON *the fish monger*

LINDA *the fish cutter*

UNSAFE AREA DO NOT ENT

DESCALER

SWORD[FISH] (500/lbs.)

KEYSTONE SPECIES[3]
A. sardines[4]
B. squid
C. herring

Fish atop the food chain (tuna, swordfish², sharks, etc) are BIO-ACCUMULATORS³ of heavy metals like mercury in their systems (it takes 30 lbs of keystone species³³ like sardines to make 1 lb of tuna), including keystone species in our diet would result in getting less heavy metals and a reduction in the overfishing of larger fish species. Over 40% of the protein fished from the world's oceans are keystone species. They are used to feed pigs and chickens (and even farmed fish.) By shifting our eating habits we can turn those fish into a healthier protein source.

Forage fish have the 'keystone species'³³ that support the entire marine food web.

1. This "food chain" is actually less a linear set of connections, with apex predators (like swordfish) at the top and algae at the bottom, than a MARINE FOOD WEB³ featuring a myriad number of complex interdependencies based on the conversion of sunlight into protein.

(Russ⁵ includes a number of recipes featuring keystone species like sardines¹ at his Oakland restaurant.)

RUSSELL MOORE'S RECIPE FOR GRILLED SARDINES WITH CHILES AND HERB SALAD 1) begin by first gutting and scaling the fish (removing the head right behind the gills), 2) rinse quickly and pat dry; 3) brush the sardines with olive oil and season inside and out with sea salt; 4) Grill over medium hot coals on both sides until crispy brown. Do not overcook! FOR SAUCE: 1) grind whole chihuacle, cascabelle or guajillo chiles (splash boiling water over chiles) 2) peanut garlic... molten and peak and add to chiles with a pinch of salt, a squeeze of lime and good olive oil. Serve with herb salad. Fish. To take better roasted on the bone.

simply fished out. We know this because examples of Barton Seaver's "object impermanence" have already happened.

"Here on the West Coast, our sardine fishery collapsed from overfishing coupled with other natural stressors," recalls Jennifer Kemmerly. "And on the East Coast we saw our cod stocks disappear because of too much fishing pressure."

A solution might come from understanding the push-pull factors at play. "When a cod net is pulled aboard," Barton Seaver explains, "up come haddock, hake, pollack, cusk, ling, monkfish, wolf, dog, skate, ray, and all sorts of stuff, yet cod is the only fish that's profitable even though they're all equally profitable to the human body for the purpose of sustaining our health."

The popular assumption is that U.S. consumers only buy ten types of fish. As a result, fishermen do whatever it takes to meet this narrow demand, even if these fisheries may collapse as a result. Attempts to establish policies and market restrictions that protect individual fish

species are complicated by a single fact: Fish respect no borders. They cross international waters, moving with impunity between the high seas and **EXCLUSIVE ECONOMIC ZONES (EEZ)** or sovereign territorial limits that extend two hundred miles beyond a country's coastline. When developing countries with weak central governments have coastlines, their EEZs often lack the capacity to enforce catch restrictions, while high-seas fisheries operating outside EEZs are simply too big to police.

It's a mess, or as Carl Safina repeatedly tells me, "It's complicated."

Since no single governing body can effectively manage global fisheries, **ILLEGAL, UNREGULATED, AND UNREPORTED (IUU)** fishing is rampant. Fishermen catch what they want, where they want. It's everyone's problem, but no one's responsibility. Call it the perfect storm. Instead of transparency, the fishing industry runs on confusion. And this negatively impacts not only our oceans, but consumers. When there's no cod or salmon, substitutions are often made. **SEAFOOD FRAUD** is a major global problem. A 2013 study by Oceana, an international marine conservancy, analyzed more than 1,200 seafood samples taken from restaurants and supermarkets across the United States. Their findings were amazing.

"About a third of what we looked at was mislabeled," Dr. Michael Hirshfield, chief scientist for Oceana, explains, "with range levels as high as almost one hundred percent for things like white tuna and not much better than that for red snapper. You ought to

> 1/3 of all fish you buy isn't what you think it is.
>
> (Research findings collected by Oceana.)

"SEAFOOD FRAUD happens any time a customer thinks he or she is getting one thing and they're getting another. It can be mislabeling, a weight lower than what's advertised, or a fish the customers thinks is caught by one particular gear [method] when it's really caught by another."
—Dr. Michael Hirshfield, chief scientist for Oceana

be able to buy fish without having a PhD, or go into a market and get what you're told you're getting."

Ironically, Carl Safina thinks the problem is less about fraud and more about the necessity of dumbing things down for consumers. "Seafood is mislabeled because most people don't have the mental bandwidth to understand all the different species we're talking about," he suggests. "Someone could say, 'We eat three kinds of fish: salmon, canned tuna, and shrimp.' Well, you just named probably fifty different kinds of sea creatures, coming from dozens of countries. People cannot sort that out, so sellers simplify it by saying something is what it really isn't.

"If you say, 'I'm just eating tuna fish,' well, tuna fish is three or four different species of tuna. There's a species called 'albacore' when it's in the can and there's 'light tuna,' which is usually two kinds mixed together, yellowfin and skipjack tuna. Salmon is at least six species of salmon, and shrimp are dozens of species, wild and farmed, from dozens of countries. You're not going to have somebody asking for a specific species of shrimp from a specific place; it's just too complicated."

Barton Seaver agrees. **"Consumers aren't educated enough to tell the difference between red snapper, mutton snapper, lane snapper, gray snapper, tilapia, catfish, or grouper. They all meld together, especially when they're sliced paper thin and served sashimi-style or nigiri-style with gently warmed rice and wasabi. We just don't know the difference."**

It may sound reductive, but Seaver sees a potential solution, one that puts the onus on consumers to get informed and ask the right questions about the fish they eat.

"We need to examine why we are so fixated on certain species," he contends. "If you can cook cod, you can cook haddock. If you can enjoy snapper at a sushi bar, you can enjoy whatever fish they're serving you. Just ask them to tell you the truth. There's no enlightenment in that. There's no extra skill set involved. It's just an honest assessment of how the industry works."

Sometimes even knowing the difference between a red snapper and a grouper isn't enough. If a fish doesn't come with **CHAIN OF CUSTODY** documentation, even an educated marine biologist can only guess at what he's eating or how long it's been out of the water. Eighty-six percent of seafood sold in America is imported. **A tuna caught in the Central Pacific might be processed in Asia, processed again in the United States, then iced in various warehouses and distribution centers before finally coming to you, the customer.** It's a long and winding road, one taken without documentation; there's often no telling where your fish has been.

"The only way a consumer can be reassured that they're getting what they paid for is with a verified chain of custody from boat to plate," observes Hirshfield. "That sounds like a big deal but it's increasingly something we're seeing for all different kinds of things. Look at conflict diamonds from Africa. People wanted to know if they were buying diamonds mined under unsavory conditions or used to finance human rights abuses. These companies quickly figured out a way to place a bar code directly on the diamond that allowed anybody to track it back to its source." Fish are next.

My alarm at Honolulu's Ala Moana hotel goes off at 3 a.m. I'm groggy, but after two coffees I'm almost ready for Scotty Fraser of Norpac, a fish wholesaler based in Seattle. The company buys catches from literally hundreds of fishing vessels working the Pacific, and Fraser is the company's man on the ground in Honolulu. We meet in the hotel lobby. Fraser is too awake for this hour. He never stops talking, but fortunately what he says is fascinating. People speak endlessly about the need to introduce **TRACEABILITY** into the fishing industry. Not Fraser. He's too busy doing it.

We start at Pier 38, site of the only fish auction between Tokyo and Maine. It's still dark when we arrive. Boats are tied to a pier. A small dockside crowd looks on as winches lift fish from the icy holds of boats that have crisscrossed the Pacific for weeks. The fishermen observe the proceedings in silence. A quick glance at this scene explains what sustainability means to global fisheries. All these vague notions of EEZs and IUUs and the Tragedy of the Commons have no place here. These fishermen are here for a paycheck. Their only concern is the size—and value—of this haul.

I wander around on deck taking pictures, distracted, still half asleep; I'm occasionally slapped by tuna and opah as they're pulled from the hold and dropped into a metal container. I notice a forklift transporting the fish to a nearby building. I follow. Inside men with cell phones circulate. These are fish buyers. They move quickly through rows of bigeye tuna set on a bed of crushed ice. Notches have been etched out of each tuna's tail fin. These, along with core samples taken from the tuna belly, are placed atop each fish. **The buyers study the flesh color, how it reflects light, then hold the samples up to their noses. An experienced buyer deduces the health and vitality of a fish, even how well it's been preserved, from such limited information.**

The auction begins soon afterward. The crowd starts at the beginning of each row, pressed in tightly as they jockey for position before each fish. The seller barks out fish weights and prices, nods to offers, coaxes these prices higher, closes deals, scribbles out sales tickets, places them on tuna, then moves down the row. It all happens very fast. There's lots of tuna to sell and

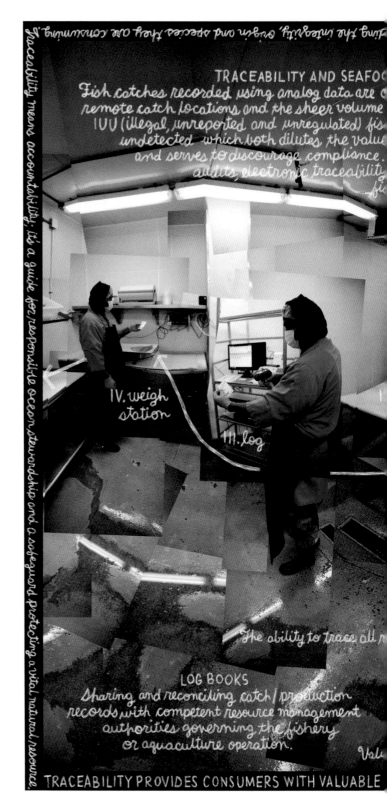

...AUD
...ifficult to verify due to
...a involved. This allows
...ter the logistics chain
...e legitimate fisheries
...implifying third party
...es the incentive to
...on compliant ways.

TRACEABILITY AND OVERFISHING
If the goal is to fully exploit a fishery based upon scientific management,
then the data used must be as complete, accurate and timely as possible.
Electronic Traceability, which is third party validated, provides that higher
level of confidence.

I. the opah

RM-ID⁼
raw material
identification
code

FG-ID
finished good
identification
code

V. vacuum
pack

VI. ready to ship

...cutters

(includes vital information about vessel,
...ecies, weight, price, origin, date, and grade)

TRACEABILITY

SCOTTY

Norpac
Honolulu, HI
13 December 2012

...rials and added components —from source through final disposition— in a logistics chain.

TRANSPARENCY
The ability to easily, economically,
and readily view the entire logistics chain.

ACCOUNTABILITY
Recording all inputs including feed and its components,
pharmaceuticals additives, ingredients, packaging that
come into contact with the raw material or finished good.

3RD PARTY VERIFICATION
...ll raw materials (origin, species inputs, disposition)

everyone's impatient. I'm jostled, elbowed, and at one point knocked to the ground. No one helps me up. They've already moved down the row to bid on the next fish.

Each tuna has a fish tag stapled to its tail. Its bar code contains information about where the fish was caught, its weight, and how long it's been out of the water. Since international fisheries operate in the absence of sufficient government oversight, some players in the seafood industry like Norpac have put their own traceability programs in place. The fishermen who sell to Norpac tag each fish as it comes out of the water. **This catch information stays with the fish when it is offloaded from a fishing vessel, cut into filets, logged,** **weighed, vacuum packed, then shipped to a restaurant or market.** The same fish. The same verified information. Traceability means **TRANSPARENCY**, independent third-party verification, and a strict accounting of the fish's journey.

Certified labels like that of the Marine Stewardship Council or lists like those made by Seafood Watch promote verifiable traceability practices. It's a start and a potential solution that couldn't come soon enough.

"I think there's incredible pressure on our fishery stocks," notes Thomas Kraft, Norpac's Seattle-based founder. "Some of that pressure comes from modern fishing practices, which harvest at much greater capacity than ever before in human history."

and Bycatch Reduction Services (BRDs). The latter is a triangle with a hoop sewn into the net so trapped fish (and sometimes shrimp) can swim back out.

(they need separate bycatch licenses for each state). In accordance with bycatch laws they use both Turtle Excluding Services (TEDS)

Louisiana, Mississippi and Alabama. Richard & Melonie shrimp throughout

Bycides, if they ate their catch they'd be eating their profit, so eating bycatch provides an economical form of sustenance.

(pickers table)→

A SHRIMPER'S BYCATCH
ground mullet, popeye mullet,
white trout and flounder.

Mississippi fishermen make
"Biloxi Bacon" from their bycatch
(they fry ground mullet in grease
drippings until it's golden brown)

Biloxi, MS
24 June 2011

BYCATCH
anything other than the target species which
ends up getting caught in a fisherman's nets
(for shrimpers this can include crab, fish and squid)

RICHARD AND MELONIE JOHNSON SAY THE MOST SUSTAINABLE THING THEY CAN DO WITH BYCATCH IS EAT IT (AND THEY DO).

Ocean acidification, different gear technologies, **BYCATCH**, changes in ocean temperature, and pollution all put additional stress on global fisheries, but the main impact on traceability is technology. Or the lack of it.

"Historically, fish have been tracked with paper and pencil," Kraft continues. "That's a very cumbersome way to work, especially when you have a lot of volume and your inventory comes and goes very rapidly. It's difficult to follow with analog traceability, but if you can bar-code a catch from a vessel or a particular fish—especially if it's big enough, like with a tuna or swordfish or even an individual salmon—you can follow it all the way from the consumer back to the fishing log that a vessel prepared when they first unloaded their catch, then you combine that with third-party observation, whether it's onboard or dockside observers, that can be either a management authority or a government or even a third-party verifier ensuring that the data recorded is accurate and correct.

"We can match each boat's catch data electronically. It legitimizes our fish and gives our customers a great assurance that what we're telling them and the facts are well aligned."

Traceability won't bring sustainability to our oceans, but the data it collects may allow us to finally quantify our consumption of ocean-based protein.

That's important because as our global population rises, more effective stewardship of these resources will be required.

Norpac is serious about implementing technology-based traceability systems. In the months after our meeting, Scotty Fraser continues our fractured conversations with phone calls from Indonesia and Vietnam, then the Philippines. These are economies with scant infrastructure investment, but their local fish processors have cast an eager eye on the massive U.S. market; they listen attentively as Fraser explains the magic and wonder of fish tags and bar-code readers. From there the conversation expands to bird lines and circle hooks, proper cold chain handling, and quality control—the sustainable methods now practiced by more evolved and transparent fisheries.

It's a process. Fraser offers these fisheries the opportunity to reach new markets, but he explains that the consumers in these markets have become more informed. They've begun to demand traceability, and have realized that through the purchase decisions they make— choosing one fish over another—they can collectively reach straight through the supply chain, traveling back to the boats themselves, to shift how the fisheries industry functions. You don't think consumers can do that just by making more informed decisions at their supermarket fish counter? Just remember the story of a cage-free egg.

MARICULTURE and AQUACULTURE

I once worked in Shanghai. After my first month, and having rarely left my seventeenth-floor office, I became restless and hired a driver to visit Wuzhen, a fabled water town outside the city. The landscape consisted of fields and nondescript marshland, but after an hour I discovered barges floating in the water. When the road rose I found myself looking down at thousands of pigs milling about in open boats, like a cockeyed, industrial-sized monoculture version of Noah's Ark. I tried asking my driver to explain the scene, but he spoke no English. His frustration was palpable.

Upon arriving at Wuzhen I arranged for my driver to wait a few hours. As it was lunchtime he pointed me in the direction of a fish restaurant. After staring uncomprehendingly at the menu for a few minutes, I rose from my chair and simply pointed at whatever dishes I saw that looked interesting. My waiter took notes, then gently slipped his hand under my arm and guided me through the kitchen. We stopped before dozens of fish tanks lining two walls of a dark room. The man gestured for me to choose one, so I did. It was orange.

My food came minutes later. The side dishes that looked spicy were spicy. I could feel each follicle inside my nose burn with heat. Then my fish came. As I was about to debone this orange mystery my driver burst into the restaurant. He had a woman with him. "I speak English please," she announced. "You want to know about the fish farm?"

"Well, no. About the boats," I clarified. "The pigs in the boats."

"Pigs live in boats, yes."

"Forever?"

"Yes. Live in boats," she replied brightly. "Eat in boats. Go to bathroom in boats. Floor open, bathroom go into water."

"And that's it?"

"Then fish eat it. Is a fish farm."

She pointed eagerly at my dish, then at the other diners.

"China is looking to get up to 60 percent of their annual seafood production from **AQUACULTURE**," observes Thomas Kraft of Norpac. "Seafood's going to be the only viable way to have protein by 2050. Even if we remove all the forests from the planet and plant them as agricultural crops, we still couldn't sustain nine billion people. **So seafood, whether it's wild-captured or AQUACULTURE, is going to play a big part in feeding the world's population."**

Fish farms are either land-based (aquaculture) or ocean-based (**MARICULTURE**). Each has its own challenges. Ocean-based fish farms are still a work in progress. Placing fish in static net pens leads chemicals and excess nutrients from feed and feces to gather on the ocean floor, killing off local fauna. Cramped conditions inside the pens, often in excess of an area's optimal **CARRYING CAPACITY**, create an ideal incubator for parasites and infectious diseases. Just as CAFOs keep cattle on a steady diet of antibiotics, these fish farms have no choice but to do the same to keep their fish healthy. Additionally, the normal flow of tides and currents in these waters not only helps spread diseases between fish within a pen, but extends pathogens to neighboring farms and even wild fish stocks. Situations are further complicated by escapes. **When fish bred for captivity escape, they often mate with wild fish, degrading the quality of these wild genetic stocks.**

As for their feed, it's chiefly made from other fish. Nearly a third of all fish caught each year are used to feed other animals. The world's largest anchoveta fishery in Peru is almost exclusively dedicated to producing fish meal that finds its way into feedstock for cattle, pigs, poultry, and other fish.

"I get a little bit nervous about finfish aquaculture because you have to start looking at the feed," Dr. Michael Hirshfield points out. "I am in favor of herbivore plant-eating aquaculture, although it always appears that once you start growing fish, you want them to grow as fast as you can, so giving them supplemental protein in the form of fish meal from

A FISH FARM
1. ample access to fresh water
2. carefully regimented feeds
3. selective breeding
4. aquatic species raised in a controlled environment
5. low density raceway which mimics its natural habit

Lake Logan

CROWDER SCREEN
pushes trout to rear end of raceway

Pigeon River

CARRYING CAPACITY

The maximum amount of fish raised in a given area of water (raceway) while maintaining a healthy environment for the species (subject to seasonal variability with less fish in warmer months, more fish in cooler months).

Sunburst Trout
Canton, NC
1 July 2011

RAINBOW TROUT
(a native American species)

PROCESSING
Within 30 minutes they are gutted, filleted, pin-boned (with individual bones removed by hand), trimmed, weighed, sized, and packed

USING FRESH STREAM WATER DIRECTLY FROM NC'S PISGAH NATIONAL FOREST MEANS THESE TROUT ARE FREE OF PCBs, PESTICIDES AND MERCURY.

other fish is what often happens. **In places like China that have a long tradition of raising species like carp, and more recently tilapia, the demand for fish meal is actually leading to more fishing in the ocean to create fish on land."**

I visit an old rainbow trout farm in North Carolina's Pisgah Forest. The men wade through chest-deep water, using wood-framed screens to push trout toward the near end of a concrete runway, where they're captured in handheld nets and loaded into rubber garbage cans for the short trip across the grounds to a single-story processing plant. Some fish tumble out of the garbage cans, flop around, then slide into a narrow drainage channel. They quickly swim away.

   "Where does this channel end up?" I ask the man.

   He points at the nearby river.

   "What happens when it gets to the river?"

   The man looks at me, puzzled, then laughs. "I suppose it's gonna make a fisherman downstream pretty happy," he replies.

The fish have nowhere to go at Will Allen's fish farm in Milwaukee, Wisconsin. They live inside wood boxes lined with plastic sheets, with the water—and waste—continually cycled through overhead containers, where it's utilized by everything from tomatoes to salad greens before returning to the water below. **Compact, holistic systems like these work and offer effective and economical ways to provide protein directly within low-income communities.** As a result, thousands of aquaculturists from around the globe trek to Milwaukee each year to learn from Will Allen.

Will

Kate

Ryan

← perch

PRODUCE
While tomatoes only grow in the summer months, over 150 varieties of produce, including spinach, arugula, chard, turnip and collard greens, lettuces, and peppers grow throughout the year.

AQUAPONICS =
aquaculture (fish farming) + hydroponics (growing plants in nutrient-enriched water instead of soil)

GP raises about 100,000 fish per year. These include tilapia, a warm-water fish native to Africa, and lake perch, a cool-water fish native to North America.

Their national outreach programs teach community leaders across America how to grow, process, market and distribute food in a sustainable manner.

GOOD FOOD REVOLUTION

Food resilience means the creation of a community food system that can reliably produce adequate good food that's safe, wholesome, and affordable to all.

PROVIDING PEOPLE FROM DIVERSE BACKGROUNDS EQUAL ACCESS TO HEALTHY, HIGH-QUALITY, SAFE, AND AFFORDABLE FOOD.

Shellfish farms are a different story. They don't need fish meal pellets, and because they're filter-feeders, they actually clean the water instead of polluting it, all of which I discover in Cape May, New Jersey.

Thousands of boats once plied their trade on Delaware Bay. Cape May Salts, a local oyster prized for its rich, firm meat, was barreled (packed) and shipped to New York, Philadelphia, Chicago, and even San Francisco. As this shellfish industry flourished, it provided the region with jobs and a steady supply of locally sourced protein. By the 1880s harvests reached nearly a million barrels a year. Overfishing inevitably ensued (remember the Tragedy of the Commons?). By the 1920s, the industry had collapsed.

Over the past twenty years, the Cape May Salts have made a modest comeback, partially due to a fertile collaboration between Atlantic Capes Fisheries and Rutgers University, which operates the Haskin Shellfish Research Laboratory in Cape May. It's an impressive facility, one focused on disease-resistant oysters specific to Delaware Bay. The project shows how universities can use science to help rebuild local food systems.

"My favorite examples of successful fish farms tend to be shellfish farms," Dr. Hirshfield concurs. **"I like the idea of mussels, clams, and oysters—which generally have been eliminated from their former habitat—being put back into the water. You don't need to feed them anything; in fact, they can reduce levels of algae when they're too high while having a minimal impact on the ecosystem as a whole."**

Casson Trenor, a seafood restaurateur who works closely with Greenpeace, agrees with this assessment of the aquaculture industry. "I think aquaculture will play an absolutely crucial role in the future of seafood products," he notes. "There's no way to get around it, but what that aquaculture looks like and what that aquaculture's impact is on the oceans and on wild animals is something that is still to be written. Aquaculture is not new. Aquaculture is thousands of years old. We've been seeing herbivorous fish raised in backyard ponds in China for three thousand years.

"If we're going to accept that aquaculture is part of the future—and I think we have to—it makes it all the more urgent to create aquaculture systems on a global level that mimic the natural and intrinsic checks and balances of a highly biodiverse and functioning ecosystem rather than operating on a completely assailable and broken paradigm, one that raises carnivores, using huge amounts of protein input for a low amount of protein output, one that creates genetic perturbation and disease vectors, and all the things that have become the norm in Western aquaculture."

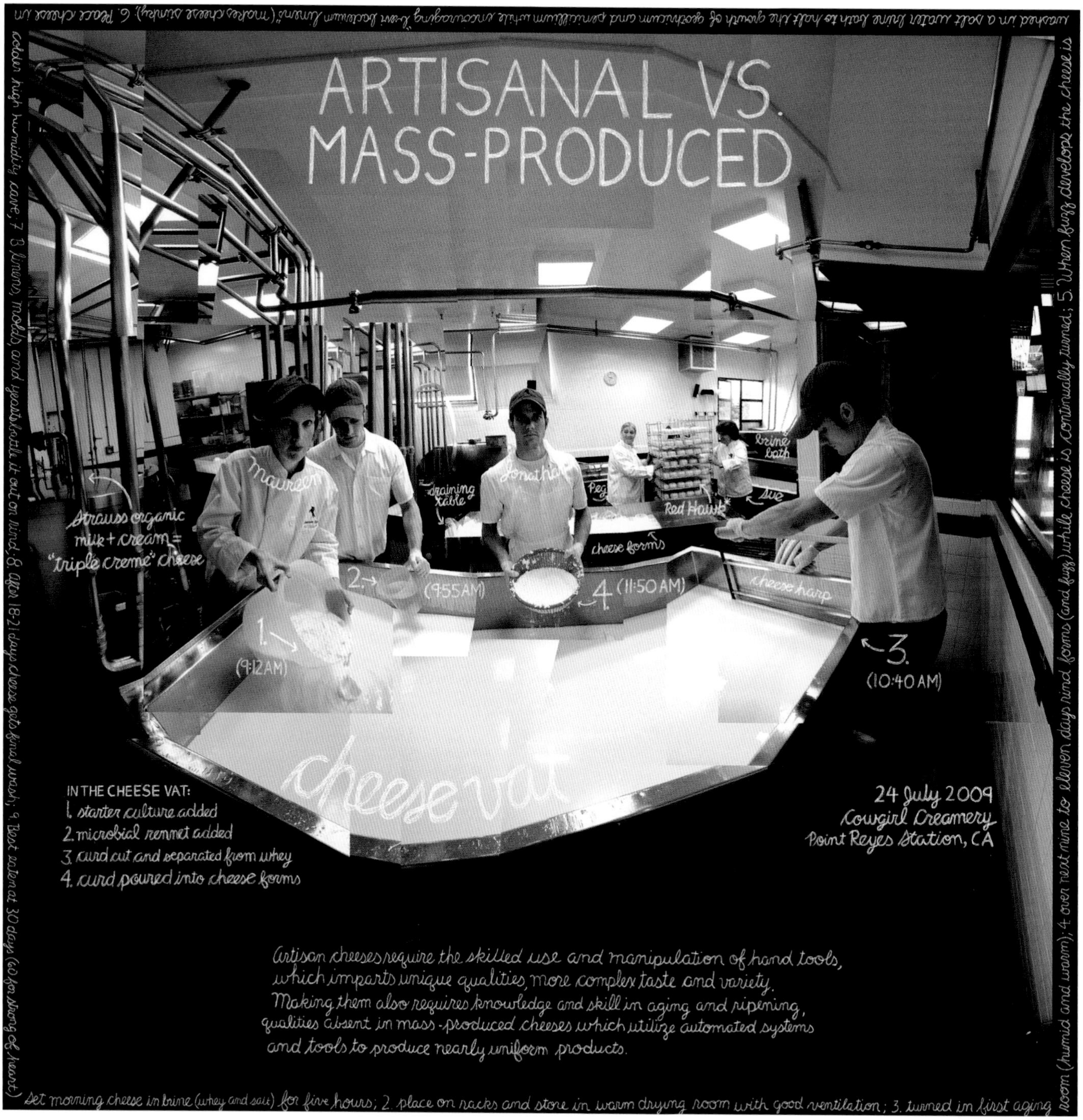

ARTISANAL VS. MASS-PRODUCED

Strauss organic
milk + cream =
"triple creme" cheese

maureen

Jonathan

Reg

Red Hawk

sue

draining table

brine bath

cheese forms

cheese harp

2→ (9:55 AM)

4 (11:50 AM)

1. (9:12 AM)

~3. (10:40 AM)

cheese vat

IN THE CHEESE VAT:
1. starter culture added
2. microbial rennet added
3. curd cut and separated from whey
4. curd poured into cheese forms

24 July 2009
Cowgirl Creamery
Point Reyes Station, CA

Artisan cheeses require the skilled use and manipulation of hand tools,
which imparts unique qualities, more complex taste and variety.
Making them also requires knowledge and skill in aging and ripening,
qualities absent in mass-produced cheeses which utilize automated systems
and tools to produce nearly uniform products.

Set morning cheese in brine (whey and salt) for five hours; 2. place on racks and store in warm drying room with good ventilation; 3. turned in first aging room (humid and warm); 4. over next nine to eleven days rind forms (and figs) while cheese is continually turned; 5. When figs develop the cheese is washed in a salt water brine bath to halt the growth of geotrichum and penicillium while encouraging brevi bacterium linens (makes cheese stinky) 6. Place cheese in colder high humidity cave. 7. 3 linens, molds, and yeasts halt it out on hand 8. after 18-21 days cheese gets hand wash; 9. first salad at 30 days (60 for young of heart)

Friends who visit my farm near Petaluma, California, are often surprised to learn that I can't tell them about the hottest new restaurants in neighboring San Francisco. I do know of a child-friendly pizza place off Geary Boulevard where the owner scans your table, assesses the situation, then orders for you. I sometimes stop at a taco truck that's always at the same parking spot in the Mission. And I still frequent a dim sum diner from my childhood—its walls covered with aging black-and-white photos featuring Miss Chinatown contestants—but that's about it. My culinary sense of San Francisco is defined more by memories than menus.

It's the same when I travel. In returning to towns I'd been in years before, I'll often retrace my steps, decoding vaguely familiar landmarks to return to that same café, that same table, and, often, that same waiter. The sense of continuity is powerful. The meal, that bottle of wine, that cup of coffee, not only exist in the present, but resonate backward through time, pulling me into a past that is not solely my own.

Which is not to say I don't appreciate good food. It's just that I'm a *foodist* rather than a *foodie*. I'm less interested in the fashionable aspects of food and more taken by food's ability to connect me to a world larger than myself. Food stripped of its cultural identity, that exists in a purely decorative, geographically neutral state, is less interesting than food imbued with a sense of place.

Food is defined by culture—losing one means losing the other. Today, parents don't know how to cook or what to buy, while children are increasingly unaware of where food comes from. And the average age of the American farmer is fifty-seven years old. We are a country that is quickly forgetting how to grow its own food. Either our national identity will be defined by box stores, microwave meals, and fast-food chains, or we'll reclaim that which defines our communities and our culture, safeguarding our culinary traditions for future generations.

THE POLITICS OF FOOD

The terms **FOOD SECURITY** and **FOOD SOVEREIGNTY** are often used interchangeably, though they mean different things.

The more conventional definition of food security is one based on calories, on having enough food to keep oneself alive. It explains whether a person knows where his or her next meal is coming from, but even this definition is incomplete. Those incarcerated in American prisons, for example, are food secure—they can expect food each day—while those living in poverty are often food insecure, meaning they may not know where or how or when they'll eat their next meal.

Wayne Roberts, former director of Toronto's Food Policy Council, provides a historical context for the term. "It comes out of American foreign policy in the 1970s," he claims. "The rising price of oil rippled through the entire economy but was most profoundly felt in the field of food. **Oil is fundamental to food production in a hundred and one ways, including the transportation of food from farmer to processor and from processor to customer, and the energy used when a customer drives to the supermarket.**"

After food prices went through the roof across southern Asia and sub-Saharan Africa, the United Nations organized the first World Food Conference in Rome. Henry Kissinger's keynote address proclaimed, "Within a decade, no child will go hungry, no family will fear for its next day's bread, and no human being's future and capacity will be stunted by malnutrition."

The concept of food security began as a consumer-focused statement, one designed to protect people from going hungry, but it made no mention of the needs of farmers or the environment. Forty years later that definition has only slightly broadened.

Kristin Reynolds, a professor at the New School for Public Engagement in New York City, notes that **"food security has now been reinterpreted in some places as** *community* **food security, as access to fresh, healthy, and affordable food not just on an individual level, but within the entire community."**

As Reynolds notes, it's not enough for a community to have access to food. It's equally important to understand where that food came from and at what cost. A culture's social stability is at stake. Both the United States and the European Union heavily subsidize food that's exported to countries in the Global South. These goods enter the marketplace with an unfair advantage. By undercutting prices set by local producers, these imports increase market share while wiping out local agrarian economies. Typically governments set tariffs to protect their markets against this flood of subsidized products, but the World Trade Organization stands in steadfast opposition to such government intervention.

"For many people, the technologies and policies offered to provide food security are very much like being in prison," author and activist Raj Patel observes. "Government will say, 'We are handing the world over to big corporations so they can produce commodity crops, then you'll have to eat them. And we'll supplement your income so you can afford these basic commodity crops.' But that's like being given a voucher for McDonald's and a bag of vitamins to cover the nutritional difference. That's not freedom."

It is here where the challenge of food security becomes one of food sovereignty, a community's right to decide how they're fed. The term was coined in 1993 by a gathering of farmworkers and small-stake food producers from around the world. Their first meeting in Mons, Belgium, led to the formation of *La Vía Campesina* ("The Peasants' Way"), which protects the rights of cultures to defend their control over local and regional food systems. The group has since grown into one of the world's largest social movements.

"You can certainly have food security under dictatorships but you can't have **FOOD SOVEREIGNTY**," Patel notes. "You need democracy for food sovereignty to happen. Food sovereignty is a deep and expansive idea that unfortunately we see too little of. Food sovereignty requires discussion. It takes putting people around the table, with meetings to figure out how water and food

are shared, and how hunger is eradicated. It looks a lot like a bunch of food policy councils. It looks a lot like kids learning at school where their food comes from. It looks like a food system free from agricultural subsidies and free from the marketing that agriculture is allowed to employ on our children. Most of all it's characterized by conversations around hunger, poverty, and community. Those kinds of conversations are happening from Detroit to Oakland and that's something to be celebrated."

This farm is surrounde
(we are right in the m
suburban housing t
New Orleans).

Xuyen Pham's Garden
East New Orleans, LA
24 June 2011

FOOD
livelih
The ability of com
independent of outside
fruits and vegetables and fishe
farmer's market, c

AFTER XUYEN PHAM LOST HER NEW ORLEAN

proved too much so they moved yet again, settling in the "Mary Queen of Vietnam" community in East New Orleans.

ed by a hotel owner in Oklahoma, but the cold

(this
is taro)

Xuyen Pham stands amidst taro plants
in her home garden. The plant stems are
a base ingredient in traditional soups
and congees found on most Vietnamese
dinner tables. By growing taro and
other vegetables she keeps Vietnamese
traditions alive in her community.

Xuyen

ouses
of a
n East

SOVEREIGNTY=
d + self-determination

y members to control food access (both effluent and influent)
urces (such as supermarkets). Its community elders grow traditional
o shrimping, fishing, and crabbing to sell at local stores, the local Saturday
st importantly, to feed their families and community members.

E TO HURRICANE KATRINA, SHE TURNED THE PROPERTY INTO A FARM TO FEED HER COMMUNITY.

She fled Vietnam with her husband and children at the end of the Vietnam War in 1975. After months in southeast

FOOD IS CULTURE

We live in a homogeneous society, a landscape of cultural sameness defined by an endless parade of big-box retailers and chain restaurants across the nation that guarantee you'll eat the same meal no matter where their restaurants are located, effectively reducing the notion of geography to a marketing campaign ("It's Louisiana Crawfish Sauce Week!" "Try our Santa Fe spice sensation!"), which is to say we are what we eat but we might be better off if we ate like our grandparents.

Most Americans are immigrants, meaning they came from somewhere else. My grandparents arrived on boats from Spain and Italy one hundred years ago. While the images of their homelands dimmed, food remained a cultural constant, but even these touchstones faded with each passing generation. Despite my best intentions, I'm proof of that. I remember my grandmother making ravioli by hand on Sunday afternoons, carefully placing each serrated square on floured trays beside her stove. Years later I make them myself, but they're not the same.

Now what passes for culture in our family is largely appropriated. In October we forage for porcini mushrooms beneath the coastal pines near Point Reyes. Our Thanksgiving table features heritage breed turkeys raised by local farm kids. December means Dungeness crab. In the summers we pick blackberries from thickets planted when we first bought the farm. Some berries we eat right there. The rest become jam. My wife has begun canning dilly beans using a recipe passed down from a great-aunt in Virginia. **We're consciously creating these new traditions not for ourselves but for our daughter; these are deposits in *her* cultural bank. Hopefully she'll draw from it when she has children of her own, but we'll only know that answer decades from now.**

Our family's condition—one of cobbling together new traditions—is hardly unique. We've become a nation of transplants. Perpetual wanderers. Most of us don't live where we grew up. Our tethers to the land, to the communities that raised us, have been severed.

Future generations will either be defined by the seasonal endcap displays marking the start of each shopping aisle at their local big-box retailer, or by the conscious act of restoring that which once defined their communities and families in ways no marketing campaign could ever hope to capture. Like everything else, it's work.

I travel to Decatur, Georgia, to meet war refugees from Burundi. An organization called the Global Growers Network helped these women start *Umurima Wi Burundi* ("The Women of Burundi Farm") at a playground just a block off East College Avenue. The women are savvy; they grow a mix of East African crops like amaranth greens (*mchicha*), maize, and sweet potatoes for their community, plus vegetables like Swiss chard, spinach, and arugula as cash crops for local customers. The farm not only helps these women keep their culture alive, but also provides them with the economic opportunity to continue with the same livelihoods they practiced in their own country.

"El Perfecto sabores cocinar con...

The durian is revered in southeast Asia for its large size, formidable thorn-covered husk and unique odor, which has led to the fruit's banishment from hotels and public transportation in Brunei, Indonesia and Malaysia.

(durian)

CRISTINA
From: São Paulo, Brazil
Business: "Kika's Treats"

She makes sweets she remembers from her childhood in Brazil.

AZALINA
From: Penang, Malaysia
Business: "Azalina's"

She teaches people about her culture through the stories behind her food.

VERO
From: Mexico
Business: "El ...

She brings classi... Mexico D.F. ... to San Francisco

La Cocina
(Mission District)
San Francisco, CA
19 January 2011

KITCHEN INCU...

Helps entrepreneurs launch, grow and formalize food businesses which can...

LA COCINA ("THE KITCHEN" IN SPANISH) PROVIDES AFFORDABLE COMMERCIAL KITCHEN SPACE AND TECHNICAL ASSISTANCE TO LOW-INCOME AND IMMIGRANT W...

ALICIA
From: Mazatlan, Sinaloa, Mexico
Business: "Tamales Los Mayas"
She translates Mayan cultural
traditions into food using
generations-old family recipes.

As Robin Chanin of Global Growers Network points out, "Children go to school and become integrated, but adults either stay home or do shift work in factories. The farm project, especially given its urban location, is an opportunity for these women to break away from that."

How well have they adjusted? They share their indigenous seeds with Burundi farmers across the United States, operate two urban farms and a **CSA** in Decatur, and the women have become a fixture on the local **FARMERS' MARKET** scene.

Caleb Zigas helps immigrant women from Asia and South America preserve their cultural identities with La Cocina, a nonprofit **KITCHEN INCUBATOR** in San Francisco's Mission District. La Cocina provides these women with the kitchen equipment and professional mentorship needed to start their own food businesses. It also allows them to transcend the cultural barriers within their immediate communities and reach a much wider marketplace in San Francisco. Their wares are now available at farmers' markets and specialty shops across the city, and a few of the women have even opened local restaurants. And the city is richer for it.

EDIBLE EDUCATION

Robbie McClam and his son Eric built City Roots farm between an asphalt emulsion plant, a railroad switching yard, and a local airport in Columbia, South Carolina. From the very beginning the 2.75-acre site presented unusual challenges. An industrial dumping accident on an adjacent lot contaminated the local groundwater, requiring the McClams to install an extensive water filtration system. The farm's first foray into aquaculture also had problems. A pump shut off during an unexpected June heat wave, wiping out their first stock of tilapia. While the McClams persevered, their commitment led to unexpected results. **Their organic urban farm, a first in Columbia, attracted a community eager to not just buy food, but learn how it's grown.** While the farm now hosts a variety of events ranging from harvest dinners to food truck rodeos, food education has become its primary function. The McClams offer both self-guided and staff-led tours, educational workshops, an internship program, and volunteer opportunities for local residents.

"We're proud to be educating the public, both young and old," Robbie says, "because we've found we are the link many people feel they've been missing in their lives. **Through our farm we've taught others not only about sustainable agriculture, but about what seasonal availability and food biodiversity mean, and their impact on local food systems.**"

Robbie's story is both special and commonplace. By necessity, farmers today are educators. Not only do they go to tremendous lengths—at much personal sacrifice—to grow wholesome food, but they also have to educate their buyers. How else will these consumers grasp the challenges farmers face in pursuing an ecologically responsible alternative to the conventional agricultural model? Suffice it to say, it's a lot of work.

K. Rashid Nuri is the founder of Truly Living Well, an organization that operates five urban farms in Atlanta, many of them built directly on concrete foundations where public housing projects once stood. **"It's not just about something to eat down here,"** he explains. **"It's about access to food that's not fast food. It's about education and employment."**

While Rashid grows food for farmers' markets, local restaurants, and a CSA, and even offers welfare recipients produce at half price, effectively doubling their food budgets, he really uses the farm to provide the community with both nutritional education and practical farming knowledge.

"We teach people how to grow food and engage the community in a process that helps them attain horticultural literacy," he notes. "And it's transformative because **sharing food with another human being is a very deep thing. It's one of the most intimate experiences you can have, which is why food is integral to the overall development of any community.**"

"Tell me and I will forget,
show me and I may remember,
involve me and I will understand."
— confucius

City Roots Farm
Columbia, SC
30 June 2011

EXPERIENTIAL
LEARNING
Education through direct involvement; offering staff-led tours,
workshops, internship programs, volunteer opportunities
and dinners to help a community engage with local food.

CITY ROOTS IS THE ONLY URBAN FARM IN SOUTH CAROLINA GROWING FOOD AND PROVIDING AN EDUCATIONAL AND EXPERIENTIAL OUTLET FOR THEIR COMMUNITY.

When people consider educating kids about food, they inevitably think about Alice Waters and her Edible Schoolyard project. Waters is a national treasure. No single individual in this country has worked so tirelessly to help Americans become more connected to the food they eat. Throughout my own journey to make the images for this book, I repeatedly heard people share their "Road to Damascus" moments, stories that explained the inciting incidents that irrevocably changed the course of their lives. More often than not, Waters played a principal role in their stories. She's a tireless visionary who

school ↑

garden →

dining hall ↓

Alice Waters started the Edible schoolyard in 1996 to educate kids about food.

JULIAN

THE SCH

Martin L

9 December 2010

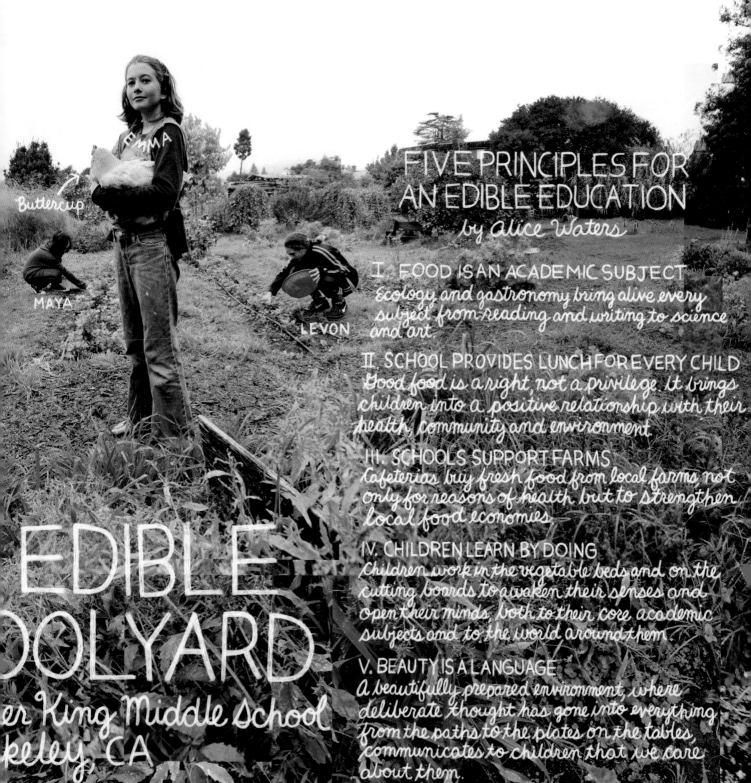

Buttercup

EMMA

MAYA

LEVON

FIVE PRINCIPLES FOR AN EDIBLE EDUCATION
by Alice Waters

I. FOOD IS AN ACADEMIC SUBJECT
Ecology and gastronomy bring alive every subject from reading and writing to science and art.

II. SCHOOL PROVIDES LUNCH FOR EVERY CHILD
Good food is a right, not a privilege. It brings children into a positive relationship with their health, community and environment.

III. SCHOOLS SUPPORT FARMS
Cafeterias buy fresh food from local farms, not only for reasons of health, but to strengthen local food economies.

IV. CHILDREN LEARN BY DOING
Children work in the vegetable beds and on the cutting boards to awaken their senses and open their minds, both to their core academic subjects and to the world around them.

V. BEAUTY IS A LANGUAGE
A beautifully prepared environment, where deliberate thought has gone into everything from the paths to the plates on the tables, communicates to children that we care about them.

EDIBLE
OOLYARD
er King Middle School
keley, CA

(They grow kale, peas, fava beans, onions, okra, tomatoes, eggplant, squash, zucchini, and herbs.)

K. RASHID

EUGENE

Truly Living Well
Center for Natural Urban Agriculture
East Point, GA

HORTICULTURAL LITERACY

Understanding who grows your food, the quality of that food, and how it gets to you is a transformative pedagogy that leads to economic development and the creation of community.

K. RASHID NURI BUILT AN URBAN FARM WHERE THERE HAD ONCE BEEN AN ATLANTA HOUSING PROJECT. "Food is integral to the overall development of any community," he says. "It is the beginning and the end. Sharing with another human being is one of the most intimate experiences you can have. But it's not just about something to eat, but a bridge. Food deserts are about more than geography; they happen when folk are scarce and education is poor. That's why TLW provides more than just quality food through its markets and CSAs. They use workshops, summer camps and a variety of other educational tools to engage the community.

simultaneously travels along high and low roads to build the necessary networks to fix our food system.

Waters was instrumental in introducing **SLOW FOOD**—an Italian initiative that combats the spread of fast-food culture by defending the security of indigenous food producers across the globe—to the United States. Members in this country quickly recast the organization to fit this country's specific cultural imperatives with a mission focused on supporting "good, clean, and fair" food. Slow Food has expanded to include two hundred chapters that organize culinary events with local farmers and guide communities in the creation of school gardens. It has emerged as the single national organization capable of defining and protecting our cultural culinary patrimony for future generations.

Meanwhile, her Edible Schoolyard Project focuses exclusively on children. Planting vegetable gardens at local schools, supporting local farmers, teaching kids how to grow and cook their own meals, and even putting their harvests on the menu at school lunchrooms all seem like fanciful concepts, yet Waters and her team have been doing this for years at Berkeley's Martin Luther King School. What started as a quixotic journey to bring food into the classroom at a single school has since become a national initiative.

Not surprisingly, Waters sees the industrial food system as her greatest enemy. "They make fast food appealing," she notes. "They make it cheap and addicting. So we really need another way to communicate with children about food."

Before Waters was a chef—she first opened Berkeley's Chez Panisse restaurant more than forty years ago—she was a Montessori teacher. **"When children are in the garden, they don't really have a sense that they're in school,"** she explains. **"Food offers a way into their young minds. It captures their attention and opens them up to different stimuli. They're smelling, tasting, seeing, and touching."**

Waters advocates getting at kids when they're still young and impressionable, and continuously strengthening not only their appreciation for real food but their connection with nature and the environment. "The public school system is our last truly democratic institution," she points out. **"Every child goes to school. What could be better than to bring them into a really positive relationship with food when they're in kindergarten? If you can touch children very early on, teach them a set of values and open up their senses to learning, you can transform the food system in America."**

Mike Todd, an enterprising high school science teacher in Ames, Iowa, wanted to show his students alternatives to industrial agriculture, so he become a curator of our Lexicon of Sustainability Pop-Up shows. These feature a collection of information artworks—many shown in this book—that provide models for **SUSTAINABLE** food production. After taking the show across the state for a year, his students began wondering how they could learn more about their own food system.

Ironically, my wife and I had already been thinking the same thing. No single photographer can document every aspect of this country's rapidly developing sustainable food system. The United States is just too big. Besides, **if we truly believe that consumers need to act locally, every community should have its own lexicon of sustainability, one uniquely suited to its people, culture, and geography.**

We conceived Project Localize to show students how to map their local food systems, identify the sustainable farming practices of local food producers, then translate these findings into information artworks to be shared with their community. It's a principle that borrows much from a Chinese proverb: "Tell me, I'll forget. Show me, I may remember. Involve me, and I'll understand." **If students travel to the places where food is grown, form relationships with these farmers, and see firsthand how demanding it is to produce socially and ethically responsible food, they'll have a much different relationship with the food they eat.**

A few months later I test the theory on a windswept prairie with seventy-five Ames high school students. They photograph prairie restoration projects, the heartbreak of soil depletion, the role of women in agriculture, and even an enterprising aquaponics project built inside what had once been a hog CAFO.

Each information artwork is a photo collage composed of twenty to one hundred images. They're large works, at times spanning ten feet across and six feet tall. Assembling them is technically complex. And that doesn't even capture the challenge of adding text, which requires a mix of journalistic skill and savvy art design. In short, making these works requires some degree of skill.

Somehow Todd convinces his students to work over the summer, and by September the images are complete. They're also amazing, just what policy makers in our nation's capital should see. We rent the entire subway station below the USDA and turn it into our a public gallery. We then decide to fly five students, along with their teacher, to Washington, D.C., for three days of whirlwind meetings with senators and congressmen on Capitol Hill.

After their plane lands, they quickly wolf down lunch, change into their finest outfits, then find themselves ushered into a windowless conference room in the Hart Senate Office Building for a meeting with Iowa's senior senator, Charles Grassley. After a few pleasantries are exchanged, the students present their work. Grassley immediately interrupts to note, "We've been farming the same way for a hundred and fifty years." The room goes quiet. Then he continues, "So I have one question for you. What about the economics?"

These kids have spent months traveling across Iowa. They've listened to stories told by dozens of sustainable farmers. They've worked diligently with us to translate these ideas into information artworks. And they've had a secret weapon: Fred Kirschenmann, from Iowa State's Leopold Center for Sustainable Agriculture, has served as their advisor. Still, how can you prepare a sixteen-year-old for a debate with a U.S. senator who's served on Capitol Hill for thirty years? What were we thinking?

But Ania Chamberlin is no typical sixteen-year-old. "Senator, the way we farm right now might make sense to you economically," she answers. "And I know a Farm Bill is currently being debated in Congress, but kids like me are more interested in the Fifty-Year Farm Bill. It's not enough that you make money from agriculture today. You're losing our topsoil and polluting our water. There has to be enough soil left for us tomorrow, for when we're adults and we need to grow food on that same land for our children. We want an agriculture that lasts. That's what good economics is."

I lean back and smile at my wife. We have a dozen more meetings over the next few days, including an audience with USDA deputy secretary Krysta Harden—who decides to hang the students' work in her offices at the USDA—and a presentation to assorted White House staff, but our work's already done. **Every revolution needs powerful distinct voices. Sometimes the most compelling are those you never factored into the equation.**

"We deserve
to know
where our
food comes from."

— WILL
11th grade
Ames High School
Ames, Iowa

"Who is
producing
your food: a
person or a
corporation?"

— CASSIE
12th grade
Ames High School
Ames, Iowa

"We should help
family farms by
shopping at our
local farmer's market;
it's healthier for us
and helps our environment!"

— TESSA
12th grade
Ames High School
Ames, Iowa

TILING

Lee Matteson

Nitrogen causes
dead zones

PROJECT LOCALIZE

57

Industrial agriculture requires fewer workers, and the fact that it's an expensive enterprise—a single combine costs hundreds of thousands of dollars—creates formidable barriers to entry for new farmers. Furthermore, the consolidation of farmland in the Midwest—another by-product of industrial agriculture—has led to the disappearance of small farms and resulted in a mass migration away from rural areas. This situation worked for a time—industrial areas certainly benefited from a surfeit of low-income workers—but now agriculture is confronted by an alarming statistic: **The average age of the American farmer is fifty-seven. What will happen when these farmers retire?**

Mark Newman is a pork producer in Missouri. In his area there were once seven hundred families making a living from pork production. "These families were able to make a sustainable income with between thirty-five to fifty sows," he recalls. "Today you couldn't do it with thirty-five hundred. We are the last pork producer in this area and actually have been the last pork producer for over ten years. **It's not that we can't find young people to get involved; it just takes too much capital and they don't have it.**"

"What about your own kids?" I ask.

"They'd be interested in coming back," he says, a slightly detectable weariness edging into his voice, "but they can make a much greater income away from the farm."

When I call Newman three months later to check in, I'm told he's dead. He'd just turned fifty-eight.

So what happens to his farm? Newman had four kids. They live hours away. The closest, from a geographic standpoint, is Chris. He lives in St. Louis and works as a contractor for FedEx Ground. He has fifteen drivers for a region that extends into Arkansas and Mississippi.

"It was never gonna be a situation where I'd just say, 'I'm going back to the farm and raise pigs in the Ozarks,'" Chris tells me. "Having been raised there and knowing what it entails . . ." His voice breaks off and the phone goes silent.

"I can see that," I reply. "It's hard."

"The things you love and the things that are practical are not often the same things," he says finally.

It's just as Newman explained when we first spoke; his children have lives of their own. Families. Jobs that pay more than they'll ever make back on the farm. And yet, Newman missed one key thing. They're still a family, even after he's gone.

"No one wants to see the whole idea die just because the leader passed away," Chris continues. **"The idea lives on in the people who helped make it what it was. That includes our family and our customers."**

And that's how their story ends, at least for now. Newman's kids adjust their lives. Rita, his widow, keeps the farm going. Everyone pitches in. The farm stays a farm.

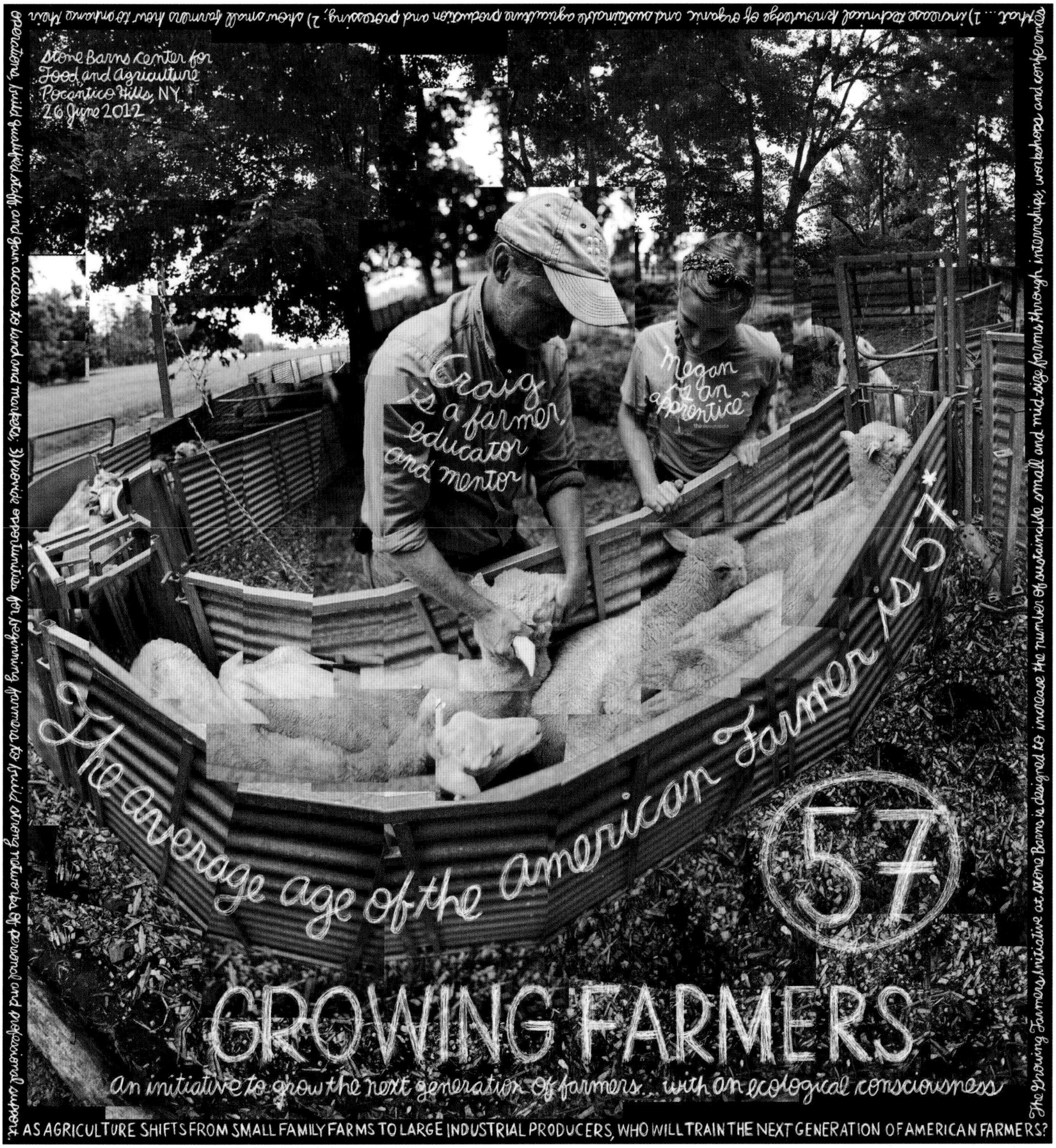

Stone Barns Center for Food and Agriculture
Pocantico Hills, NY
26 June 2012
Craig is a farmer, educator and mentor
Megan is an apprentice
The average age of the American Farmer is 57*
57
GROWING FARMERS
an initiative to grow the next generation of farmers... with an ecological consciousness
AS AGRICULTURE SHIFTS FROM SMALL FAMILY FARMS TO LARGE INDUSTRIAL PRODUCERS, WHO WILL TRAIN THE NEXT GENERATION OF AMERICAN FARMERS?
The Growing Farmers Initiative at Stone Barns is designed to increase the number of sustainable small and mid-size farms through internships, workshops and conferences
that... 1) increase technical knowledge of organic and sustainable agriculture production and processing, 2) show small farmers how to enhance their
operations, build qualified staffs and gain access to land and markets; 3) provide opportunities for beginning farmers to build strong networks of personal and professional support.

ALBA (Agriculture and Land-Based Training Program) is a **FARM INCUBATOR** based in Salinas, California. It helps farmworkers mostly trained in conventional agriculture make the leap to becoming certified organic farmers operating their own businesses. It's a program with an interesting premise: who better to understand the perils of industrial agriculture than those who've toiled in its fields?

I spend a week on ALBA's two properties; its fields are a colorful patchwork of small plots worked by farmers growing a variety of market crops from strawberries to spinach to tomatoes. The farmers here are a close-knit group, shy at first, then eager to share their stories. The fact that they've got a good deal isn't lost on them, though each is quick to tell me farming is hard work.

The program begins by providing farmers with a half-acre plot. They're given tools, access to water, and professional guidance ranging from accounting to organic certification, and are even provided with a market for their goods. Each year these farmers are granted more land and greater autonomy. After a few years in the program their rent increases. Soon their rents are higher than market rate, making it cheaper for them to venture out on their own. They leave ALBA, freeing up land for the next generation of organic farmers. Graduates of the program have become fixtures in the Northern California farming scene.

The **GREENHORNS** are a mostly informal collection of young, brainy, energetic world changers. They make films, write books, host workshops, and somehow find time to farm. While there are Greenhorns in every state, the group is headquartered in a converted Main Street storefront in Hudson, New York. On the afternoon I visit, the troops are deep into their latest project, reviving the tradition of publishing annual farming almanacs. Their mix of retro graphics, homespun farming tips, and essays on late-twentieth-century farm labor practices are a fertile collage of exuberant pastoral boosterism.

I find Megan Schilling, a self-professed **GREENHORN**, at the Stone Barns Center, a teaching farm just one hour outside New York City. A recent college graduate, she stands amid Dorsets and Finnsheep, cataloging their behavior with copious detail in a dog-eared notebook. Schilling doesn't strike me as a typical summer intern padding her résumé with a litany of character-building skills. For her, this stuff is real.

A year later I track Schilling down in Los Angeles; she works at a goat dairy and interns at a public school's **BIODYNAMIC** garden. As I suspected, it's all part of her master plan; she eventually plans to have her own farm.

"One of the most important things I've learned is that farming is totally possible," she tells me. **"As long as you have someone to teach you and a community that can help out or give advice."**

Schilling tells me she has no plans to start a farm on the West Coast. Too many water access issues. Instead she's thinking of the East Coast, in fact not far from Stone Barns.

"You know, it can be challenging, not only to find the farmland but then get the money to buy it," I point out.

"I don't need a lot of money to start up," she replies. "Still, getting everything together and putting the infrastructure in place does seem really daunting."

"A **FARM INCUBATOR** is the place where you can make mistakes only to correct them and avoid them in the future. It's a trial-and-error environment, under professional guidance, where if you fail on a particular part of your growing process you don't lose everything. Whereas, outside of the incubator, you may lose that and more."
—Tony Serrano, general manager of ALBA Organics

BANNER FROM A GREENHORN EVENT IN MAINE
It included four massage tables, herbalists, health care advocates, film screening (with panel discussion afterward) and Maine-made root beer

MITHEREEN FARM

OPEN
BUY MERCH

EXIT

WE ARE
GREENHORNS
Our mission is to
RECRUIT PROMOTE & SUPPORT YOUNG FARMERS

The Greenhorns use signs like these at events to make passersby think, "Who are those strong, young, confident people with dirty clothes?"

TO MAIN ST.
HUDSON, NY

AG LITERATURE TABLE
Community resources for those who wish to farm locally:

1. farm education courses
2. agricultural mediation
3. agricultural census
4. massage
5. grant programs
6. agricultural politics campaigns
7. mailing lists
8. bicycle shops

*(their event formula = pie + beer + workshops + ice cream + performances + spit roasted pig + dances.)

"GOVERNMENT POLICY SETS THE FRAMEWORK, BUT FARMING IS ABOUT CULTURE AS WELL AS COMMERCE. IT'S ABC

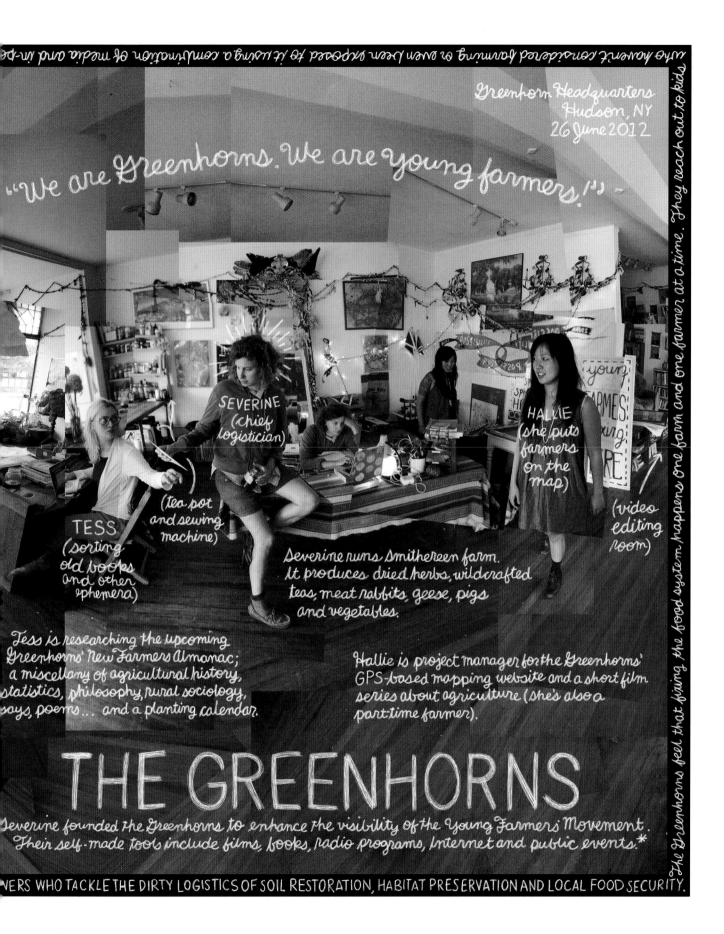

"We are Greenhorns. We are young farmers!")~

Greenhorn Headquarters
Hudson, NY
26 June 2012

who haven't considered farming or even been exposed to it; making a combination of media and in-per...

They reach out to kids. one farmer at a time. happens one farm and ...

TESS
(sorting
old books
and other
ephemera)

SEVERINE
(chief
logistician)

(tea pot
and sewing
machine)

HALLIE
(she puts
farmers
on the
map)

(video
editing
room)

Severine runs Smithereen farm.
It produces dried herbs, wildcrafted
teas, meat rabbits, geese, pigs
and vegetables.

Tess is researching the upcoming
Greenhorns' New Farmers Almanac;
a miscellany of agricultural history,
statistics, philosophy, rural sociology,
...ays, poems... and a planting calendar.

Hallie is project manager for the Greenhorns'
GPS-based mapping website and a short film
series about agriculture (she's also a
part-time farmer).

THE GREENHORNS

Severine founded the Greenhorns to enhance the visibility of the Young Farmers' Movement.
Their self-made tools include films, books, radio programs, Internet and public events.*

The Greenhorns feel that fixing the food system...

...VERS WHO TACKLE THE DIRTY LOGISTICS OF SOIL RESTORATION, HABITAT PRESERVATION AND LOCAL FOOD SECURITY.

When the money fell out of the domestic real estate market after the crash of 2008, speculative investment moved into rural landholdings, creating historic spikes in agricultural real estate prices. It couldn't have come at a worse time for prospective farmers. They were simply priced out of the marketplace.

Somehow Jeff Broadie and Kasey White got lucky. I sit cross-legged on their living room floor in Eugene, Oregon, watching them bag beans for customers making a CSA pickup later that day. They didn't have wealthy families to bankroll their dream, and their repeated attempts to secure loans from local financial institutions were unsurprisingly fruitless. Unless the farmland had a "McMansion" on it, banks weren't interested. Even worse, the young farmers were saddled with student loans from college.

"We always said we'd get a farm but it wouldn't be through the normal channels," White observes. **"It sounds cheesy, but when you love to do something as much as we love to farm, the universe sometimes steps up and provides. Magic happens."**

In their case, magic took the form of two **FARM FAIRIES**. Instead of investing in mutual funds and financial derivatives that had no connection to their daily lives, Jerry and Janet Russell made a radical decision. They pulled their money out of Wall Street and decided to bank instead on their local food system. They connected young farmers with valuable farmland, then provided the capital and financial stability needed to secure bank loans. Building out these farms and creating successful agricultural businesses was left to the young farmers themselves. It's a long-term commitment for all parties concerned.

Jump-starting small farms in a community seems well-meaning, but is it good business, especially when these farms aren't growing high-volume, big-market crops?

"When you look at the economic viability of certain investments," I ask John Bloom, a social investor with the Rudolf Steiner Foundation, "don't monocultures offer investors very clear benefits since they utilize economies of scale? I mean, that's a fundamental strategy that investors normally associate with successful businesses. Bigger crops. Greater yields. More profits."

"I put it in the framework of risk management," Bloom explains. "A whole farm organism—with multiple crops rotating through its fields and healthy, **SUSTAINABLE** farming practices—is more likely to survive the destruction of one crop than a farm practicing monoculture. **We recognize that agriculture is whole system. It isn't just about inputs and the outputs. It's about the degree to which farmers build SOIL FERTILITY, because if farmers are building soil fertility, they're going to be there for a long time. You could still boil it down to risk management, but we're looking at it from a values perspective. That's actually more important than whether they're getting twenty bushels per acre."**

Investing in land, people, and values. It's a concept echoed later by Wayne Roberts. "It's not just about consumers," he tells me. "It's also about farmers. It's not just about farmers and consumers. It's also about nature."

FARM FAIRIES

Individuals who structure creative financial agreements which allow farmers to purchase land, set down permanent roots, and dig what they love instead of being priced out of the land market by speculators.

Jerry and Janet said it was hard to find good farmers, even harder than finding good farm land. They want to encourage young farmers and be certain local good continues to be produced in their area because farmers are beginning to go extinct.

Massey-Ferguson 231
38 Horse Power (1997)

Lonesome Whistle Farm
(future home of)
Republic of Eugene, Oregon
28 February 2011

Janet Jerry Kasey Jeff field

...chickens of the world collapse, we'll still have good water, good soil, good food and community. Plus Janet likes chickens

...greater access for people who want to farm. We all believe that good farms are going to go up in value, maybe not financially, but definitely in use-value. If all the financial...

KASEY AND JEFF FELT THEY COULD TRUST JERRY AND JANET BECAUSE "THEY KNEW WHERE WE WERE COMING FROM. WE WERE FARMERS WITHOUT LAND SECURITY."

They also shared our concern about the state of the world. They understood the serious need for more local farms and

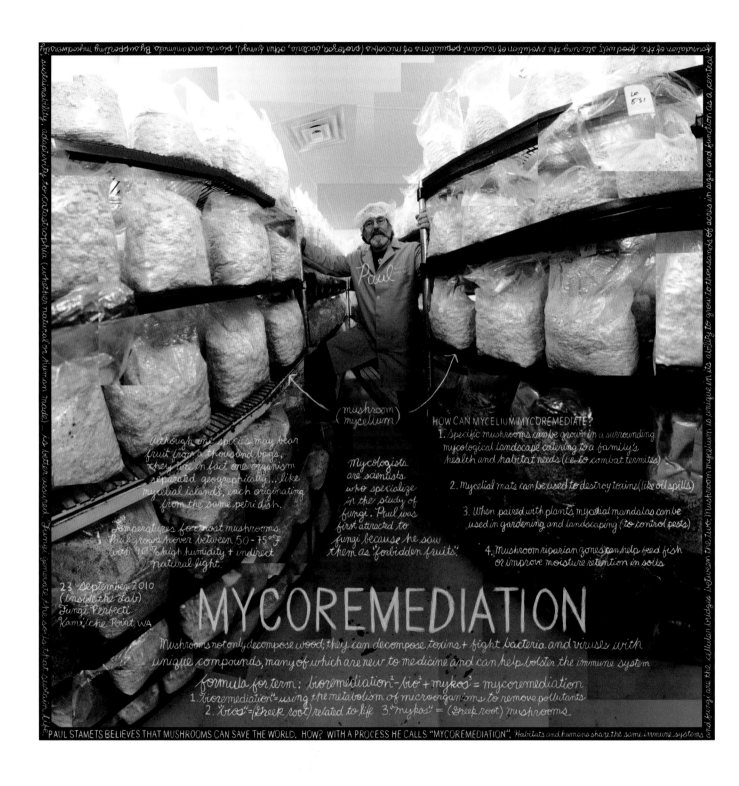

By unraveling the evolution of modern populations of nuclei-a (protozoa, bacteria, other fungi), plants and animals, foundations of the food web securing the... Humongous the mushroom-thread mycelium

sustainability, adaptivity to catastrophia (whether natural or human made)... to better around fungi generate the soil that sustain life

and fungi are the cellular bridges between the two. Mushroom mycelium is unique in its ability to grow to thousands of acres in size, and function as a central

Paul

(mushroom) mycelium

Although one species may bear fruit from a thousand bags, they are in fact one organism separated geographically... like mycelial islands, each originating from the same petri dish.

Temperatures for most mushrooms Paul grows hover between 50-75°F with 90% high humidity + indirect natural light.

23 September 2010
(Inside the Lab)
Fungi Perfecti
Kamilche Point, WA

Mycologists are scientists who specialize in the study of fungi. Paul was first attracted to fungi because he saw them as "forbidden fruits".

HOW CAN MYCELIUM MYCOREMEDIATE?
1. Specific mushrooms can be grown in a surrounding mycological landscape catering to a family's health and habitat needs (i.e. to combat termites)

2. Mycelial mats can be used to destroy toxins (like oil spills)

3. When paired with plants mycelial mandalas can be used in gardening and landscaping (to control pests)

4. Mushroom riparian zones can help feed fish or improve moisture retention in soils

MYCOREMEDIATION

Mushrooms not only decompose wood; they can decompose toxins + fight bacteria and viruses with unique compounds, many of which are new to medicine and can help bolster the immune system

formula for term: bioremediation[1] - bio[2] + mykos[3] = mycoremediation
1. "bioremediation" = using the metabolism of microorganisms to remove pollutants
2. "bios" = (Greek root) related to life   3. "mykos" = (Greek root) mushrooms

PAUL STAMETS BELIEVES THAT MUSHROOMS CAN SAVE THE WORLD. HOW? WITH A PROCESS HE CALLS "MYCOREMEDIATION". Habitats and humans share the same immune systems

PART VII

THE NEW FOOD ECONOMY

Basic economics teaches us that supply feeds demand and that producers grow what consumers want. For the past fifty years consumers have wanted cheap food, and the food industry has proven to be staggeringly efficient in meeting that challenge. Food prices have steadily fallen and food is now treated as a commodity; consumers seek out good deals and mainly buy on price.

But the cost of this cheap food is an illusion. Producers often maintain their low prices by using practices that externalize costs normally factored into the price of food. The result? Our food system is opaque; it prevents consumers from understanding how their food system works and why their cheap food isn't really cheap.

Principles like true cost accounting look at the full social and environmental impact of the myriad external costs created in producing the goods on your supermarket shelf. They provide consumers with greater transparency about their food system. The result is that many have decided to opt out of the "food as commodity" model entirely. Instead, they seek out products more closely aligned with their values. And because supply feeds demand, new markets for organically and sustainably grown foods have appeared; these producers have recognized that shifting their production practices to become more responsible stewards of their land and water resources makes good business sense. In turn, they expect to be rewarded by the marketplace, even if their newfound sustainable practices result in higher retail prices.

When the full weight of external costs associated with industrial food production is understood, principles like true cost accounting may help explain why a dozen eggs from pasture-raised chickens are closer in true price to a dozen eggs from a factory farm. Getting consumers to pay more for real pastured eggs, organic produce, or grass-fed beef becomes the next challenge. It requires that consumers be informed, and that they weight their own value systems and the societal benefits behind one farming practice versus another. Only then will they become full members of the New Food Economy.

THE REAL COST OF CHEAP FOOD

Many Americans now buy their food from the same store that sells them tube socks and lawn chairs. These consumers are loyal to familiar brands and make purchases based on price. But cheap food comes at a cost, one that's hidden from most consumers. Getting them to understand the consequences of their purchases is perhaps the greatest obstacle to fixing our food system.

In 1792, Alexander Hamilton helped found the Society for Establishing Useful Manufactures. It constructed the nation's first industrial park at the Great Falls of New Jersey's Passaic River. A succession of industries utilized the river's embedded energy to build everything from textiles to steel, Colt pistols to locomotives. These companies capitalized on nature. They depended on energy generated by the Passaic and saved money by dumping waste directly into the river. Externalizing these costs led to lower prices and greater profits, at least for a time.

Those companies have since vanished, but the **EXTERNAL COSTS** associated with the production of their goods have left a lasting legacy; the river is now one of the most polluted in America, transformed from a symbol of American entrepreneurial spirit into a Superfund site.

The price of cheap goods is an illusion. Everything costs money. You pay at the checkout stand or you pay in other, more oblique ways. These external costs are hard to see. In fact, sometimes they're only visible to future generations. Just ask anyone living in central New Jersey.

"We're now spending nine or ten percent of our gross income on food where we used to spend between twenty and thirty percent," Fred Kirschenmann points out. "We also used to spend seven or eight percent on health care. We now spend eighteen percent. You can't say it's entirely because of our diet, but certainly we know that a major portion of our increased health costs are due to the food we eat."

Why is food nowadays so much cheaper? Economists would have us believe that the consolidation of the food industry led to greater cost efficiencies and reduced

waste, but that's not true. The food we eat is mostly cheap because it's a commodity grown in a "completely rigged economic system," explains Patrick Holden, director of the United Kingdom's Sustainable Food Trust. "If you grow food in a way which exploits natural capital, diminishes soil fertility, causes emissions that lead to climate change, pulverizes biodiversity, and causes rain forest destruction, you don't pay for any of that damage," he observes. "However, farmers who deliver positive benefits to health and the environment, who create jobs, who reduce emissions while building soil carbon, who produce food while coexisting with biodiversity, get no financial reward."

He continues, "This is a world of obvious absurdities. Imagine a children's story where food companies put short-term profit ahead of public and environmental health, contaminating virtually the entire planet's food system, with those responsible getting away with it scot-free, whilst the good guys, producing healthy food and healing all the damage, don't get paid for this and go out of business. A cautionary tale, you might think, but not something that could happen in reality. No, this is actually a true story of the economic system which has produced our food for the past fifty years."

The **TRAGEDY OF THE COMMONS** explains the principle that farmers, when granted access to shared pastureland, will graze their animals as much as possible, at the expense of not only the field but other farmers, until nothing is left.

Most consumers buy solely based on price. They don't care whether food is **LOCAL**, **NON-GMO**, or **ORGANICALLY CERTIFIED**. If it's more expensive, they think they can't afford it. Consumers who want to shift their buying habits to foods that are sustainably grown need to know the real price of food, but that requires establishing a price for these external costs. If they can't be seen—or measured—it's as if they don't exist. **Therefore, consumers who buy the cheapest food in the supermarket unwittingly do the greatest social, economic, and environmental harm.** Call it the **TRAGEDY OF THE SUPERMARKET**.

WATER = EMBEDDED ENERGY

RIVER = DUMPING GROUND
FOR INDUSTRIAL WASTE

3. This is the East Coast's second largest waterfall (after Niagara Falls)

1(these costs may include pollutants released in waterways or air)

EXTERNAL COSTS[1]
The social and environmental costs excluded from a producer's retail price

(why cheap food really isn't)

STEAM LOCOMOTIVES

THINGS ONCE MANUFACTURED HERE
(but not anymore)

COLT PISTOLS

TEXTILES

2(An uncontrolled or abandoned place where hazardous waste is located, possibly affecting local ecosystems)

Allied Textile Plant
Great Falls of the Passaic River
Patterson, NJ
2 November 2012

abandoned textile mill

THE NATION'S FIRST INDUSTRIAL PARK, BUILT AT THE GREAT FALLS[3] OF NEW JERSEY'S PASSAIC RIVER, WHICH IS NOW AN EPA SUPER FUND SITE.

[left margin] years. These industries abandoned the area, leaving behind heavily-polluted waterways that are still being cleaned up.

[top margin] noted that harnessed the Passaic River's embedded energy and powered a variety of manufacturing concerns after 200

[right margin] for establishing useful manufactures" (S.U.M.). The group built a

[bottom right margin] In 1792, Alexander Hamilton helped found "The Society

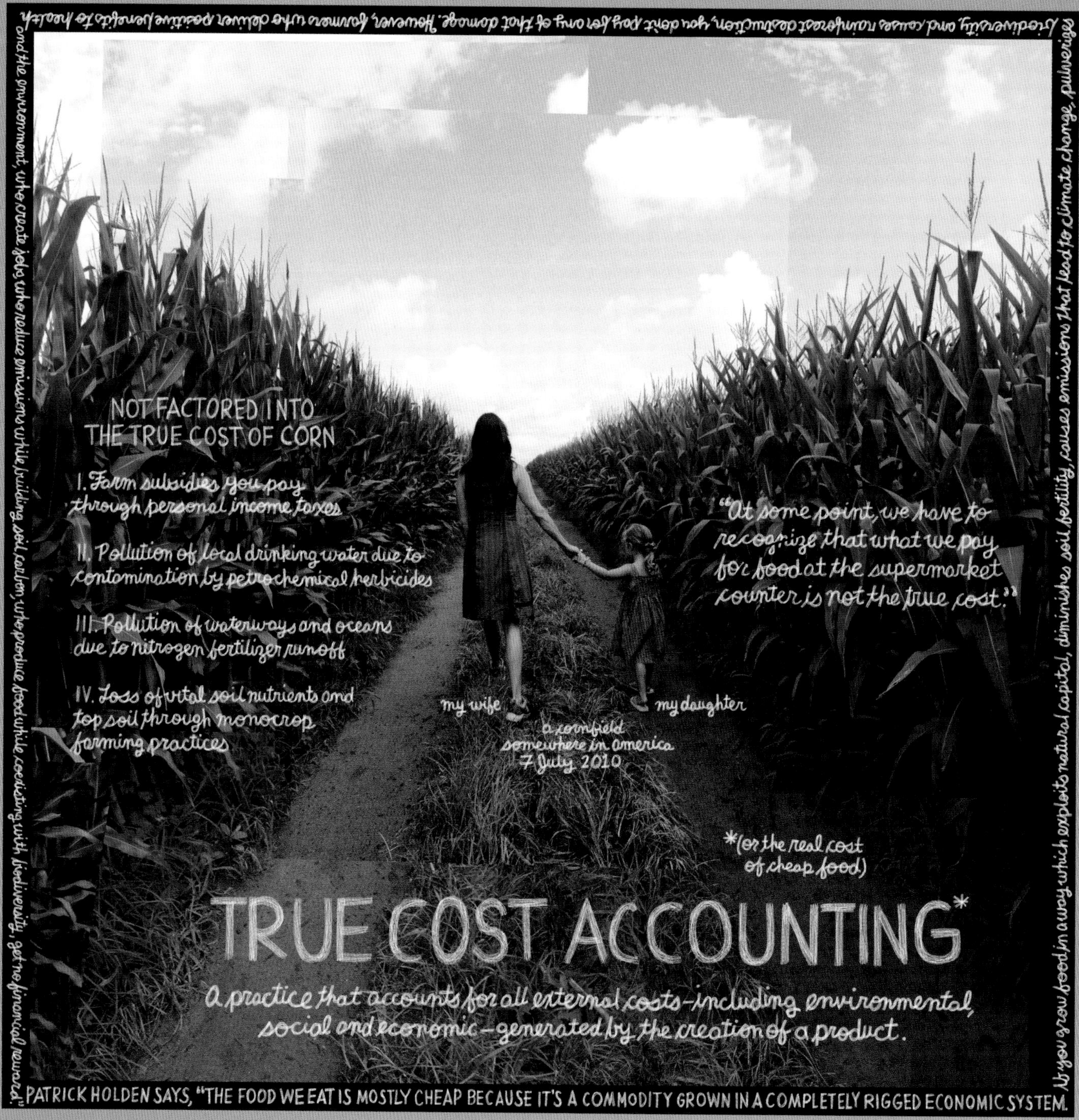

NOT FACTORED INTO
THE TRUE COST OF CORN
I. Farm subsidies you pay
through personal income taxes
II. Pollution of local drinking water due to
contamination by petrochemical herbicides
III. Pollution of waterways and oceans
due to nitrogen fertilizer runoff
IV. Loss of vital soil nutrients and
top soil through monocrop
farming practices
my wife
my daughter
a cornfield
somewhere in america
7 July 2010
"At some point, we have to
recognize that what we pay
for food at the supermarket
counter is not the true cost."
*(or the real cost
of cheap food)
TRUE COST ACCOUNTING*
a practice that accounts for all external costs—including environmental,
social and economic—generated by the creation of a product.
PATRICK HOLDEN SAYS, "THE FOOD WE EAT IS MOSTLY CHEAP BECAUSE IT'S A COMMODITY GROWN IN A COMPLETELY RIGGED ECONOMIC SYSTEM.
If you grow food in a way which exploits natural capital, diminishes soil fertility, causes emissions that lead to climate change, pulverizes
biodiversity, and causes rainforest destruction, you don't pay for any of that damage. However, farmers who deliver positive benefits to health
and the environment, who create jobs, who reduce emissions while building soil carbon, who produce food while coexisting with biodiversity, get no financial reward."

"At some point, we have to recognize that what we pay for food at the supermarket counter is not the **TRUE COST**," Fred Kirschenmann notes, "but determining the true cost of cheap food will be difficult given the food industry's lack of transparency." If we had data on every disruption in the food chain—the cost of cleaning up polluted water, the health of farmworkers exposed to dangerous conditions, or simply the loss of rural communities triggered by consolidation in the agricultural industry—**LIFE-CYCLE ASSESSMENTS** would let us "internalize" these external costs, thereby making it possible to compare and contrast the true cost of every product in our food system.

It would also make clear that we live in a world where two economies operate in parallel. The first is **EXTRACTIVE**. It depletes our natural resources for the sake of production, taking without putting back. Conventional agriculture would be an example of an extractive industry. Farmers grow single-commodity crops that depend on fertilizers and pesticides while depleting topsoil.

The second type of economy is **GENERATIVE**. In this economy people work together to conserve and even generate resources. They create, restore, and sustain. They build community.

"That's the future we're moving toward," predicts Kirschenmann. "We'll probably go through a painful transition—some of us will continue to insist on using the extractive economy to enhance our own personal wealth—but when the ecosystem services of our natural communities and the social services of our communities can no longer support us, it's going to become increasingly dysfunctional."

Moving toward a generative economy is a long-term process, one that first requires consumers to kick the "cheap food habit" and avert the Tragedy of the Supermarket by recognizing the impact their purchases have on both their health and the environment.

"It's a tough issue," George Siemon, CEO of Organic Valley, a farmer-owned dairy cooperative based in La Farge, Wisconsin, contends. "You watch people buy a Coca-Cola that costs a huge amount per ounce, then they say, 'I can't afford organic milk.' You can try talking to these people, but at some point they have to *want* to listen, because when they look at organic milk, they just see it compared to conventional milk. They don't compare it to Coca-Cola or the other things they spend money on."

Wayne Roberts, former director of Toronto's Food Policy Council, claims that the noble idea of cheap food was born out of misplaced priorities on a national level. "Europe made a much better decision than North America and England," Roberts notes. "They decided the thing to subsidize was affordable housing. In England and North America we decided to subsidize food. Cheap food. **The original purpose of cheap food may have been humane: to make sure people on low incomes could have enough food. We now know the road to hell is paved with good intentions. The poor are the real sufferers from cheap food in the world. They get the wrong signal from the marketplace: Food is cheap so you should buy *lots* of it.**"

Are government-subsidized food systems designed to help poor people actually hurting them more than helping them?

"People on low incomes think they need cheap food because they see themselves as consumers," Roberts continues. "The reality is that the largest occupation in the world is farming: farmworkers, processing workers, people driving trucks. Waitressing is the number-one service job in almost every North American city. None of these people benefit from cheap food; they are its victims. They all suffer because they're paid the lowest wages."

According to David Aylward, a senior advisor of Global Health and Technology at Ashoka, a Washington, D.C.–based incubator for social entrepreneurs, the food industry has just been doing its job. Aylward asserts, "We asked the food industry to meet two standards: more and cheaper. They responded extremely well. They produced lots more food—the global famines we expected in the sixties never happened—and costs went way down."

But the story doesn't end here. While the Green Revolution helped popularize the use of petrochemicals in conventional agriculture, resulting in higher yields and lower food costs, it also delivered unintended consequences.

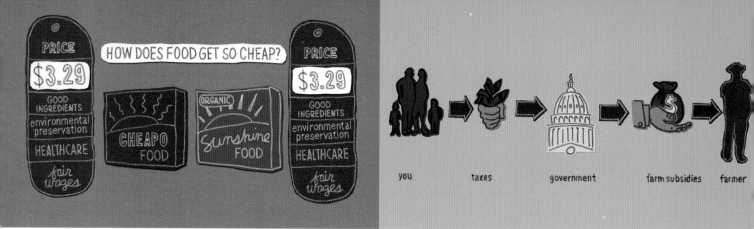

"The problem is that this food is not very good for us," Aylward continues. "It's not nearly as nourishing as it used to be, so we have widespread malnourishment and obesity—or what some might call 'over-nourishment.' While we have intervened in the marketplace to produce more at lower costs, we have not intervened in the marketplace to encourage full nourishment."

In Aylward's mind **every shortcut taken—using petrochemical inputs instead of building soil fertility, or storing unripe produce for weeks, for example, before sending it to market—translates into a critical loss of nutrients.** If technology can detect these losses, even trace them back to their sources, the food industry would change its practices. By Aylward's thinking, when presented a choice, consumers will always opt for the most nutritious food.

I ask him if there's a technology that would allow consumers to perform a **NUTRIENT ACCOUNTING** on their own food.

"If you sit on top of the Mars Rover, you'll see it uses one arm to pick up a rock and crush it," he explains. "Another arm uses a mini mass spectrometer to tell you all the elements of that rock. That same technology is now used to determine the presence of iron, retinol or vitamin A, zinc, folic acid, and iodine, the five leading nutrients that public health people look for in blood. For food you want to measure an even fuller spectrum of nutrients, but the beauty of mini mass spectrometers is that they can do that."

After Japan's Fukushima Daiichi nuclear plant meltdown in 2011, consumers in Korean grocery stores began carrying pocket-sized Geiger counters to monitor radioactivity in the fish they were buying. **It's not inconceivable that consumers may eventually carry pocket-sized mini mass spectrometers to measure the nutritional content of meat, fish, or poultry, or even fresh produce. It's science fiction today, but so was the concept of cell phones before their introduction twenty years ago.**

If consumers have the technology—or simply the information—to help choose a carrot based solely on its nutritional value, or compare the benefits between different foods, the marketplace will respond, just as we've seen with **CAGE-FREE EGGS** and **rBST** milk. Consumers, when provided with tools and knowledge that make the food industry more transparent, will make choices more closely aligned with their values, and force industries to shift their practices.

Greater transparency will also erode another aspect of the traditional consumer shopping experience. "The goal of the modern food economy is to get consumers hooked on brands and to develop brand loyalties," says Ken Cook, executive director of the Washington, D.C.–based Environmental Working Group. "For decades consumers walked down the supermarket aisle, encountering brand after brand, and only seeking out the ones they were loyal to. What we're seeing now is a different kind of habit-forming among consumers; they're starting to ask questions. They don't necessarily seek out their favorite brand. Instead, they ask themselves, 'Didn't I read somewhere that this food might lead to concerns about allergies in my family?' 'Didn't I read somewhere that if I buy this food I'm going to be contributing to environmental pollution?'

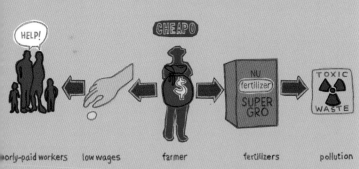

"People are making big changes in what they eat, why they eat it, and the connections they make to companies," he continues. "As a result we're going to see some pretty significant changes in how food is grown. It's a very exciting time. I don't know if we can completely change the economy through changes in our food purchases, but we can certainly change what's offered in the grocery store and the underlying value that's embodied in those products, which will increasingly skew in the direction of fewer additives, with food that much closer to organic, local, fresher, and simpler. It's going to take a monumental effort, but I think we're already seeing some indication that it's happening."

Many of the changes Cook describes will come when consumers can actually witness tangible changes resulting from their purchasing decisions. Danielle Nierenberg, cofounder of Food Tank, a Chicago-based food advocacy group, believes getting people to engage may be simpler than you realize. "People are craving for a way to connect," she says. "We need to capitalize on that craving to be involved, to feel like you're part of something in a world where you hear all these terrible things, where you read about climate change and conflict and famine and obesity epidemics. Here's a way that you, you personally, by your choices, can really make a difference. There's a real need to think of agriculture not as an industry and not food as a commodity, but as a holistic landscape. The more consumers feel that connection, that they're part of a food system, that they're not separate from it, they make a real difference."

GLEANING

Volunteers collect leftover fruits and vegetables after a field has b
with the food usually donated to school lunch programs, non-profit

IZZY

ANNA

JANE

SYDNEY

Marin Organic Glean Team
17 August 2009
Star Route Farm
Bolinas, CA

harvested,
food banks

CHESTER

spinach ↓

FOOD WASTE

Forty. That's the percentage of food in this country that never gets eaten, or that's grown and never comes to market. It's the food we distribute that never reaches a destination or sits on grocery store shelves without finding a consumer. And it's food consumers buy but never eat. It's called **FOOD WASTE**.

Sometimes at harvest, farmers have to leave perfectly good produce behind in their fields—if it's ugly, or too big or too small, it will be discarded—even though it's perfectly edible. Why? Because consumers want food that is perfect.

That's where **GLEANERS** come in. Farmers let them go onto their fields to collect leftover fruits and vegetables, with the food usually donated to school lunch programs, nonprofits, and food banks. **By performing FOOD RESCUES, communities can help redistribute food that would otherwise be wasted, helping to turn food insecurity into food security.**

Consumers can also **COMPOST**, turning food waste into valuable nutrients that improve their soil and feed their plants. Even large farms compost. Full Circle, a farm in Carnation, Washington, mixes coffee chaff from local roasters, dairy and horse manure, unused vegetables from their farm, and even organic compostable material from people like you to revitalize their soil and grow fresh produce for families across the Pacific Northwest.

Butchers and restaurateurs also work together to minimize food waste. Instead of using only selected cuts they expand their menus to include the whole animal, eating from **NOSE TO TAIL**, because respect for an animal means trying not to waste it.

And you can do your part, too. The average American throws away twenty pounds of food a month, so maybe a few things your mother told you when you were young were true: "Take what you want, but eat what you take," and "Finish your food, because somewhere, someone's hungry."

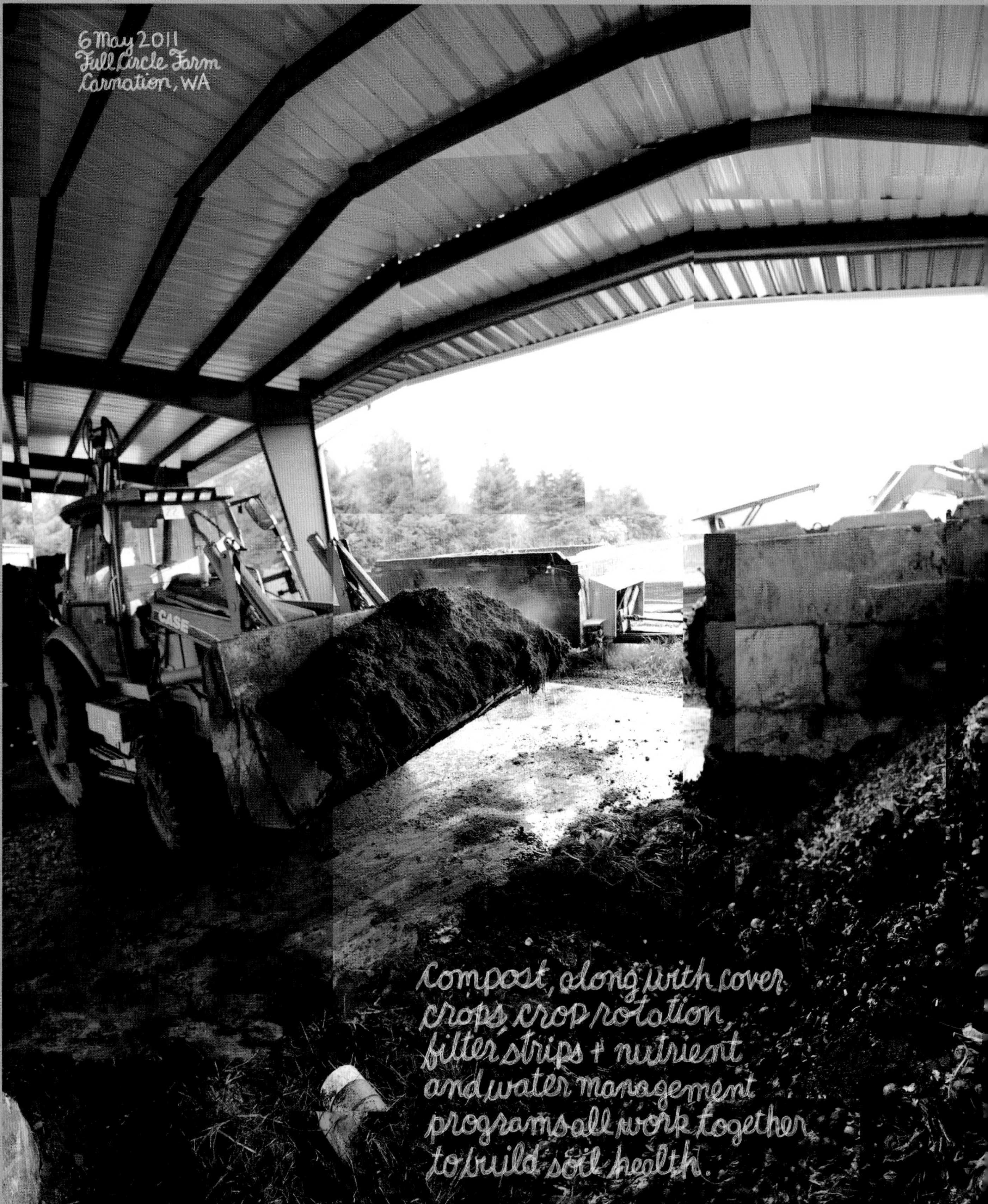
6 May 2011
Full Circle Farm
Carnation, WA
CASE
Compost, along with cover
crops, crop rotation,
filter strips + nutrient
and water management
programs all work together
to build soil health.

Jessica

Andrew

WHAT'S IN THIS COMPOST?
1. unused vegetables directly from field (vary seasonally)
2. coffee chaff (byproduct of coffee roasting)
3. dairy manure
4. horse manure from local stable mixed with sawdust or straw
5. organic compostable material (from neighbors like you!)

150°

The pile heats up to 131-150°
for one week, then heat loving
bacteria start their dance to
degrade the organic matter in
the pile and create humus.
The process takes 2-3 months.

WHY COMPOST?

mpost is a great recycler. It takes "waste materials" and creates a necessary and vital resource: top soil

1. adds nutrients back into the soil that plants utilize for 2-3 years
2. increases organic matter and promotes beneficial organisms
3. improves soil structure eroded by wind, water + tilling
4. reduces dependence on petroleum-based fertilizers

WHAT WE CAN DO TO MINIMIZE SOIL EROSION: grow crops which cover the soil during more months of the year and grow crops which add organic matter back into the soil to preserve soil quality.

FERTILIZER
Mainly composed of nitrogen + phosphorous

(This is a rain field)

HERBICIDE
GMO corn is designed for use with glyphosate

— HERBICIDE + FERTILIZER

ALICE

SOIL DEPLETION

TOP SOIL
It takes nature up to 500 years to create a single inch.

TOP SOIL + FERTILIZER + HERBICIDE

WHERE IT'S GOING
The Gulf of Mexico

The removal of nutrients, biological diversity or structural quality due to improper or extractive practices

ALICE SHOWS ME THE IRREVOCABLE LOSS OF TOP SOIL ON HER NEIGHBOR'S LAND AFTER RECORD RAINFALL IN MAY 2013.

Mustard Seed Community Farm
Ames, Iowa
28 May 2013

By following the proper principles, rainwater can be captured, re-directed, and stored for future use.

WATER STEWARDSHIP

I gather with farmers in an Iowa cornfield on one particularly wet day in May. We watch with great dismay as the most expensive and fertile agricultural land on earth—top soil that takes up to five hundred years to build a single inch—is carried off by a widening stream that etches itself deep into the cornfield before disappearing into a nearby creek. The brackish churning flume moves quickly away. It will eventually reach the Mississippi River, carrying with it a nasty brew of whatever fertilizers and pesticides convey this field's crops through a typical season of conventional agriculture.

This will inevitably clog the waterway, impeding commercial barge traffic on the river and requiring emergency dredging operations. As for the pesticides, and fertilizers carried with it, they're joined in the Mississippi River by runoff from factory farms that store their manure in giant ponds that often leak into our waterways. This sewage is particularly nasty as it often carries antibiotics and harmful bacteria. This noxious cocktail of fertilizers, pesticides and factory farm runoff ends up in the Gulf of Mexico where it creates **DEAD ZONES** that often equal the size of New Jersey. These kill fish and have become yet another challenge facing fishermen who have already dealt with hurricanes and oil spills. **Who would ever think that growing corn in Iowa or raising pigs in Missouri actually kills fish in the Gulf of Mexico?** It's a case of cause being effectively disconnected from effect, and yet another powerful example of external costs.

After I spend time with a farmer named Mike Peroni in Curtis, Washington, he writes me a letter. It begins with a description of a heavy storm one evening in May 2007. A river lining the back edge of his farm rose. By sunrise it broke its banks.

"Everything that I had to define my existence on the material plane was gone in a matter of hours," Peroni writes. **"My family was safe but I was left with no point of reference for who I was. My equipment was ruined, my vehicles, my home, the land on which we farmed, clothes, appliances, everything. Gone."**

"Humility has not figured heavily in my life," he continues. **"The flood changed all that. I was sitting in our local Grange hall with a few other folks, hunched over a bowl of soup with half a sandwich in my hand, when I started to cry. These men and women had come to the Grange every morning to make breakfast and stayed all day until the last plates where cleaned up after dinner. Why? A sense of obligation; a natural instinct to help those in their community who needed a hand. I had heard of such things. I could understand the concept and the principle at work intellectually, but I'd never felt it. I had spent the better part of my adult life unaware that this type of behavior was natural to most people.**

"A few days later a man pulled into my drive in a tilt-bed truck with an excavator on the back. He went to work moving mud out of our driveway and into a pile that we could move later. He ran for eight or nine hours without pause. When he finished he loaded up his equipment and began to pull out of the driveway. I offered him cash, but he wouldn't take it. I offered him a roast out of the freezer. He wouldn't take that either. I finally insisted that he take some vouchers that the local fire department had handed out to at least replace the fuel he used that day. He looked down at me from the cab of his truck and said, 'Don't f--- with my blessing.' Then he drove away.

"This is how blatant things had to get for me before I could become aware that people are innately good. That they are naturally kind. That in a healthy community adversity brings people together and doesn't tear them apart. So with this newfound indebtedness, and a long-overdue sense of duty and obligation, I became aware of all the opportunities the farm had to enhance our environment, set an example for other small farms, and involve the community, especially children."

What happened next is almost unbelievable, except that there's a picture of it, so you know it happened.

The flood of 2007 was a cautionary tale: Clear-cut logging triggered rain soaked mountains to dump water and silt into the South Fork of the Chehalis River. Meanwhile, farmers downstream were eager to plant as much arable land as possible. By the time Peroni bought his farm, there were no trees on either side of the Chehalis. Peroni's decision to "enhance our environment" took the form of replanting a **RIPARIAN BUFFER ZONE** with native trees on nearly a third of the farm, basically removing the land from farm use.

He didn't plant the trees himself. Instead he got help from a local elementary school. "You haven't lived until you've worked side by side with 140 fourth graders," he writes. "They slogged through the mud, endured torrential rains, and met the task with a willingness and enthusiasm that was nothing short of inspirational. I was moved to tears, both inspired by their willingness and enthusiasm and delighted with this unfamiliar sense that I was contributing positively to their future. As a farm we were providing the opportunity for these kids to have a tangible positive impact on their environment, and to visit a farm in their community where food is grown and sold directly to their families and neighbors.

"The irony, that I worked so hard to protect a river that had arguably destroyed my life, does not escape me. The flood was a natural event; no doubt it was influenced by poor logging practices and even poorer regulations regarding those practices, but the reality is I owe this river. I probably won't live long enough to enjoy the forest these trees produce, but I delight in knowing my daughter will, and I remain hopeful that some of these children, as a result of coming to the farm, might consider a future in agriculture or conservation."

NATALINA

NATIVE TREES RE-INTRODUCED

Grand Fir, Hemlock, Oregon White Oak, Western Red Cedar, Big Leaf Maple, Crabapple, Oregon Ash, Black Cottonwood, Red-Osier Dogwood, Hooker Willow, Beaked Hazelnut, Cascara

(It will take generations for this forest to establish itself. Mike and Heidi might not live to enjoy it, but their daughter will.)

HEIDI

The Great Flood of 2007 buried Mike and He...

...ed. It awakened Mike to the damage caused by generations of poor logging practices which took their toll, ravaging the delicate ecological balance of Boistfort Valley.

(it rains approximately 50 inches and 162 days per year)

Chehalis River

BUFFER

24 April 2010
Boistfort Valley Farm,
Curtis, WA

ED

MIKE

(This project was made possible by Growing Places Farm and Energy Park and ESD 13 who obtained a "Pioneers in Conservation" grant from the Washington State Conservation Commission, and funding from the American Farmland Trust and the National Fish and Wildlife Foundation).

RIPARIAN BUFFER ZONE

"The Ribbon of Life" is a vegetated area near a stream where water and earth meet. After almost losing everything to a flood, Mike decided to return his riverbanks to their natural state:

1. control erosion
2. provide shade and lower stream temperature (salmon and steelhead are cool water species at risk above 64°)
3. safeguard against flooding
4. provide habitat for indigenous species

...ily farm and home under tons of silt. Everything which defined their existence on the material plane was gone in a matter of hours. Everything. Gone. (Some of their equipment was found miles downstream.) With the

...half of Washington. The Perus closed their farm of debris and started over. The process of recovery to

What does responsible water stewardship look like? Prior to Western contact, Hawaiians governed their islands according to the *ahupua'a* system, which defined community boundaries according to the natural flow of water. These **WATERSHEDS** started on mountain ridges, then followed the flow of gravity-fed streams down, and even into, the ocean. **Just how important was water to the Hawaiian people? Their word for water is** *wai*. **For wealth, it's** *wai wai*.

Stacy Sproat-Beck is the executive director of Kauai's Waipa Foundation, which stewards one of the islands' last remaining *ahupua'a*. "In ancient times," she notes, "it was very important for us to protect our watersheds because in Hawaii—and probably with native people everywhere—the land and the resources where you lived are what sustained you."

Agriculture consumes almost 80 percent of the world's fresh water, so sustainable water stewardship is critically important, yet most people couldn't even name their local watershed. Water, like sewage treatment and electricity, has become mostly invisible. We don't know where it comes from. That's a problem, because if we don't know where our water comes from, we probably won't know to miss it when it's gone. And the fact is, our freshwater supplies aren't infinite.

Agriculture draws much of its water from aquifers, which are bodies of water beneath the ground. Nearly a third of all agricultural lands in the United States—where we grow the bulk of our corn, soy, and wheat—rely on the Ogallala, an aquifer that spans the Great Plains region. This is **FOSSIL WATER**. By some estimates it's water that's twenty thousand years old.

"The Ogallala Aquifer is a great example of a nonrenewable resource," says Peter Gleick, cofounder of the Pacific Institute, a Berkeley-based environmental research center. "We're pumping it faster than it's **RECHARGING**. Groundwater levels are dropping and farmers who depend on it for irrigation are now falling back on rent because they can't afford to pump anymore. I think it's possible to use groundwater resources in a **SUSTAINABLE** fashion, but it means for some of those resources, like the Ogallala, like with parts of northern

India or China, or in the Central Valley of California, that we have to cut our pumping rates back to a level where recharge and extraction are in **WATER BALANCE**."

When it was built in the 1930s, the Hoover Dam was the most expensive engineering project in U.S. history. It blocked the Colorado River to create Lake Mead; the result was a dependable water source that now feeds California's Imperial Valley, the largest single producer of our nation's winter vegetables. Here too **WATER MINING** has created considerable problems.

"The Colorado River is another great example of man's ability to completely consume a renewable resource," Gleick continues. "No water reaches the mouth of the Colorado River anymore. It goes to grow food in Arizona and the Imperial Valley of Southern California and alfalfa and cotton in the desert. Basically, it's a good example of our ability to do things with water that maybe we shouldn't do. If we could figure out how to rethink our management of the Colorado River and grow food more efficiently, we have a chance of not only restoring flows in the Colorado River, but changing the way we allocate water in a scarce environment."

Water is a shared asset. It belongs to all of us. Its misuse is yet another example of the Tragedy of the Commons. **As long as there's water in our reservoirs and aquifers, farmers will find a way to pump it onto their fields. If people don't understand the vital nature of protecting these shared water resources, what hope do we have for saving them?**

One idea is to institute better **WATER PRICING**.

"Water is not realistically priced," Jay Famiglietti, director of the University of California, Irvine's Center for Hydrologic Modeling, points out. "In most places it's nearly free. Maybe we pay for some treatment or transport costs, but its value as an input in industrial food processes is undervalued, so we tend to waste it. Whether we're talking about agricultural pricing or domestic water pricing or municipal or industrial use, **we need to really rethink the price structure for water, because when things are free, they don't get the respect they should in terms of using the resource wisely.**"

One way to solve the problems created by rampant misuse of water, as Famiglietti suggests, might be to create market mechanisms and impose pricing structures that motivate people to be more responsible about their water use. But even this has critics, as David Beckman, executive director of the Pisces Foundation, points out.

"Water is a human need," he says. "It's essential. Without water, humans, animals, and plants can't live. So it's true that one reason water is wasted is because its price is out of balance with its value. It's too cheap for the amount of time and effort and impact it has on a farm or business or home.

"On the other hand, since we need water, there's got to be some amount available to everybody that is not priced for profit or priced at unreasonable levels, because nobody can do without water."

A possible solution is tiered-pricing models, with the biggest consumers paying the highest prices for their water use. Another, more sustainable option would be shifting to farming practices that employ more efficient water practices.

Back in the 1930s, Simcha Blass, an Israeli water engineer, observed an interesting phenomenon when visiting a friend in the Israeli town of Petach Tikva: Among a line of trees, one was exceptionally bigger and taller than the rest. He then noticed that a coupling in a nearby water pipe was leaking, drop by drop, beside the tree. At the surface there was just a small stain of wetness, but when he dug into the soil he discovered that the wet area became wider, in the shape of a bulb; it both stored and provided water to the roots. Years later he refined his observation and discovered that by using smaller hoses and reducing water flow to mere drips, he could pinpoint precisely where to send water, dramatically reducing waste. **DRIP IRRIGATION became the greatest tool for water conservation in twentieth-century agriculture.**

I travel to Israel's Negev Desert to visit Netafim, the pioneering firm behind drip irrigation and now a global leader in water conservation practices. Naty Barak, the firm's chief sustainability officer, explains how, during the winter months, European and American markets increasingly depend on produce grown in mile after

mile of Israeli nethouses outfitted with an amazing collection of drippers remotely controlled by smart-phones. I see bell peppers and tomatoes. Cucumbers. Onions. It's incredible that so much production has been coaxed out of such an inhospitable place. Similar growing practices that maximize water efficiency in favorable winter climes are now under development in northern Mexico, California's Imperial Valley, and Arizona. Hyperefficient systems like drip irrigation will have increasing importance as water mining continues to outstrip the rate of replenishment at both Lake Mead and the Ogallala Aquifer.

Consumers who monitor their own **WATER FOOTPRINT** by taking shorter showers or turning off the bathroom faucet when brushing their teeth might do even more by simply changing what they eat. One example is meat consumption.

As Sandra Postel, founder of the Global Water Policy Project, notes, "We've had a tripling of the global population since 1950, but meat consumption has increased six times, so it's been increasing twice as fast as population."

"It's often said that when the Chinese middle class discovers the hamburger, we're finished," I point out. "We simply don't have enough cows or grazing land to support that much demand."

"Red meat is a very water-intensive part of our diet," she explains, "especially if the meat is grown in the typical fashion we've seen for the last several decades, with cows fed grains in a feedlot. It takes a lot of water to grow that grain so—calorie for calorie—meat is very water-intensive compared to most other things we eat. A very simple action, like eating a bit less red meat, can save a lot of water."

Peter Gleick agrees. "The less meat we eat, the less water is required to grow our diet," he points out, "but it's more than just what we eat. We've gotten used to thinking, 'Gee, I can have those blueberries in the winter. All I have to do is import them from somewhere faraway.' But that takes water and energy resources. Part of our sustainability challenge is to get tuned in to the total implications of the choices we make as individuals from both a water and an energy perspective."

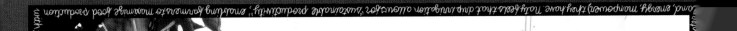

"Moses got water from a stone" says Naty Barak of Netafim, "But he forgot to share this precious technology with us, so we had to invent drip irrigation"

Ibrahim manages Ein Hatzeva's water operations (and he does it using his smart phone)

DRIP IRRIGATION

A system of plastic tubing with sophisticated drippers spaced at a set distance that enables the slow, precise and targeted application of water and nutrients to a specific location at the root of the plant in a way which maximizes water utilization while preventing water evaporation, runoff and waste.

WHY USE NET HOUSES?
1) expand the growing season on a farm
2) reduce moisture loss to evaporation
3) protect against invasive plant species and pests to reduce need for pesticides
4) reduce + control radiation for some crops

Moshav Ein Hatzeva
Central Arava Valley, Israel
14 December 2012

↑
INSIDE:
organic red bell peppers

Farmers in Israel's Negev Desert deal with water scarcity, high soil salinity and harsh climate conditions.

LAND STEWARDSHIP

In some parts of the country, a farm is worth more for its real estate than for what it grows. And as profits become increasingly elusive, some farmers look to sell off their land. Their farms become housing developments or parking lots for corporate industrial parks. In many parts of the country, valuable, food-producing land, which has been carefully tended for generations, has been lost forever.

Bob Berner recognizes the value of preserving the undeveloped open space and agricultural traditions in a community. He directs the Marin Agricultural Land Trust in Northern California's Marin County, which helps keep farmers and ranchers on their lands, ranchers like the La Franchi family. Four generations of La Franchis have continuously operated their dairy in Nicasio, California. From grandmother to grandson. Cousin to cousin. Daughter to mother. Wife to husband. Son to brother and brother.

A **LAND TRUST** helps preserve this ranching family by creating a **CONSERVATION EASEMENT**. How does it work? **The trust purchases the development rights to a piece of farmland to ensure it stays in its current use.** In exchange, these ranchers receive cash, which they can use to invest in improvements, expand operations, pay off debts, or, as in the case of Orchard Creek farm in Lansing, North Carolina, a conservation easement helps ensure that a blueberry patch stays a blueberry patch. As one of the farm's owners, Walter Clark, observes, "A conservation easement creates cultural continuity in an otherwise unpredictable world."

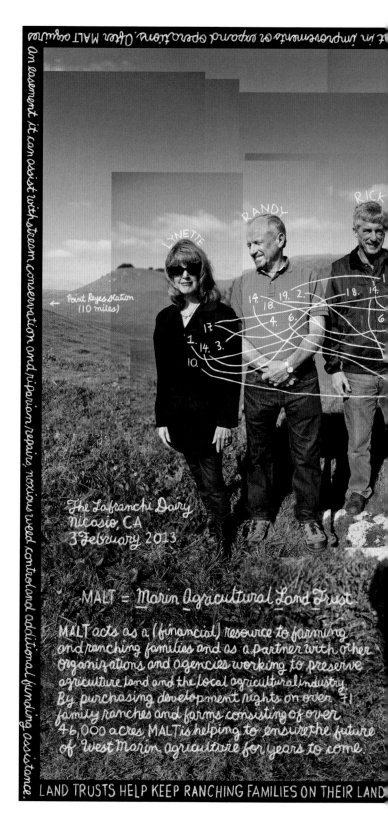

An easement it can assist with stream conservation and riparian/repair, noxious weed control and additional funding assistance. [...] an improvements or expand operations. Often MALT acquires

LYNETTE RANDY RICK

← Point Reyes Station (10 miles)

The Lafranchi Dairy
Nicasio, CA
3 February 2013

MALT = Marin Agricultural Land Trust

MALT acts as a (financial) resource to farming and ranching families and as a partner with other organizations and agencies working to preserve agriculture land and the local agricultural industry. By purchasing development rights on over 71 family ranches and farms consisting of over 46,000 acres MALT is helping to ensure the future of West Marin agriculture for years to come.

LAND TRUSTS HELP KEEP RANCHING FAMILIES ON THEIR LAND

KAREN

3 1
10 13

MARY

SCOTT

NATALIE

18
19
6 2 4 14

RICK JR.

12 8

5.
15.

9 6
12.
16.

KENDALL

15
8

milking parlor

San Francisco
(32 miles)

*27 more family
members aren't
in this picture

4 GENERATIONS OF LAFRANCHIS*
1. mother 2. father 3. wife 4. husband
5. daughter 6. son 7. grandmother
8. granddaughter 9. grandson
10. daughter-in-law 11. mother-in-law
12. cousin 13. aunt 14. uncle 15. niece
16. nephew 17. sister-in-law
18. brother-in-law 19. brother

BOB BERNER
Executive Director
MALT

LAND TRUST

A private nonprofit organization that actively works to conserve
land by undertaking or assisting in land or conservation
acquisition and by its stewardship of such land or easements.

MALT offers two types of assistance : The purchase of an agricultural conservation easement, which enables the owner/family

LAFRANCHI BELIEVES THAT WITHOUT MALT'S VISION, WEST MARIN WOULD HAVE LOST ITS AGRICULTURAL BASE LONG AGO.

wood sorrel

RUNNING SQUIRREL
Cherokee Forager

near Dougan Falls
Skamania County, WA

SUSTAINABILITY

Respect Mother Earth. Respect the land. Learn from the animals.
When foraging always leave something behind for whoever comes next.
In this way you're sure to find something when you come back.

WHEN THE INDIANS CAME UPON A NEW LAND AND DIDN'T KNOW WHAT TO EAT, THEY ASKED MOTHER EARTH FOR GUIDANCE.

"If only she picks their leaves are taken, a plant will regrow. If cut off at the ground, it dies. Our people, they want the whole plant."

WILD SALAD: Lettuce and wood sorrel. Rabbits like licorice root. Running Squirrel mixes it all together for his famous root.

lemon balm, mustard and wood sorrel (for a "dry taste") + miners lettuce, wood violets and chickweed (for "moistness").

Her answer was to eat what the animals eat: Elk eat ferns, skunk cabbage and licorice root. Deer search for miners

WHAT IS SUSTAINABILITY?

I'm told to seek out a legendary Pacific Northwest forager, an old Cherokee named Running Squirrel. He has neither a phone nor an email address so it takes me weeks to track him down. Our meeting is finally set up through an intermediary: Wednesday morning at ten o'clock in a Burger King parking lot in Camas, Washington. I arrive on time, sit in my car fifteen minutes, get out, survey the parking lot, then go inside the Burger King and ask every male over the age of fifty if he's Running Squirrel. After receiving puzzling and not altogether pleasant denials, I go back outside and wait another ten minutes. I nearly convince myself that a gentleman staring at me from a booth inside Burger King is indeed Running Squirrel, despite his earlier claims to the contrary. I am about to go back inside and confirm my speculation when I notice a red sedan parked across the road. The driver's window is down and I see a man behind the wheel.

I walk over and ask if he's Running Squirrel. The man nods, eyeing me carefully. Running Squirrel, it turns out, practices the fine art of observation.

An hour later we move through the dense understory of a forest beside the Washougal River. He gingerly lifts stones to see what lies beneath, searches at the base of fallen trees, scans riverbanks. He's foraged these lands between the Washougal and Mount St. Helens since early childhood, having first learned the craft from his mother and an aunt. They sent him

out each day with a simple admonition, he explains. "Watch the animals. See what they eat. Then you'll know what you can eat."

We stop in a small clearing. He removes a bag from his pocket and starts picking wood sorrel. It's a small leafy plant that vaguely tastes of sour lemons. "They're not anybody's plants," he observes, scanning beyond the clearing. "They're Mother Earth's. She put them here for our consumption. Like she told the Indians in the beginning, 'Don't take everything. Leave some for seed.' I believe that's what you call . . ." He stumbles, unable to pronounce the word.

"Sustainable?" I offer.

"Yeah, " he mutters, almost to himself. **"Once you take every fish out of the river there's nothing left to harvest. It's gone. Don't forget tomorrow."**

Running Squirrel's principle of sustainability is one of land stewardship. Don't consume everything. Leave a little behind so you'll have something tomorrow. It's a notion that recalls something I mentioned earlier in this book, a definition attributed to Canada's Commission on Conservation, which explained back in 1915 that "each generation is entitled to the interest on the natural capital, but the principle should be handed on unimpaired."

When applied to our food system, the concept of sustainability depends on who you talk to. It's either a term that's been rendered meaningless by marketing overexposure or the key to our survival.

FOOD

She grows all her vegetables throughout the year, including cole crops each winter (brussels sprouts, cabbage, cauliflower, collards, kale, and broccoli). She also raises her own meat (rabbits and pigs) which she slaughters on the property.

WATER

Ann's water for both irrigation and drinking comes from the spring in a nearby pasture (she collects rainwater to wash her dishes and do laundry)

cracklings

sauerkraut

canned carrots

pickled banana peppers

rendered pork fat (lard)

ANN

KATIE

(Ann cans and preserves food throughout the year)

Ann's home atop Rose Mountain Lansing, NC 2 July 2011

DIY

Do It Yourself

Don't rely on the grocery store, the mechanic or the firewood delivery truck; being self-reliant and sustainable means learning to do everything yourself (instead of laying out hard earned cash for someone to do things for you)

knife sharpener

sausage grinder

coffee grinder

apple slicer/peeler

sieve

↑ ANN'S APPLIANCES

FARM FRESH EGGS

where their food comes from need to be educated. They must re-learn the value of independence, reliability, self-assuredness and freedom."

(inaccurate) relationship with its land and food is all wrong. Only farmers have a relationship that lends them to the earth. Those who don't look to

"People are scared to live with less because they don't know how to do or anything with their hands except use a computer," she says. That's why

"ANN ROSE HAS LIVED BY HERSELF ATOP THIS REMOTE MOUNTAINTOP FOR EIGHTEEN YEARS, UTILIZING OFF-THE-GRID SKILLS LEARNED FROM HER GRANDPARENTS."

Methane from cow waste is a greenhouse gas and one of the largest single forms of air pollution in California. It's literally heating up the planet. My local dairy decided to do something about it. They installed a methane digester that converts methane waste into a useful fuel to power machinery on the farm, reducing their energy costs by 90 percent. **That's how sustainability works: a series of small, deliberate gestures that both inspire people and shift a society's consciousness, as we become local.**

Another local farmer sells me heritage breed turkeys each year. She grows them in the shadow of a tumbledown farmhouse that has been in her family for five generations. One day, while making my yearly November farm visit for turkey, I learn that her father recently passed away, leaving her the farm. The barns are now full of rabbit cages. The back field has sheep. These are her new businesses. **"Sustainability is also survivability," she explains. "It matters how I farm, but it also matters that this farm not go away."**

It's a theme I hear again on Bayou Sauvage outside New Orleans. In 2005, a Gulf Coast fisherman named Peter Gerica lost everything to Hurricane Katrina. The storm surge destroyed his property and left Peter in a tree with only the clothes on his back. He rebuilt the house—this time on stilts—and got new boats in time to experience the BP oil spill in 2010. For Peter, **sustainability means survivability as well, but it's also a tribute to his indomitable spirit.**

I also discover the same on a North Carolina mountaintop. It's early July but Ann Rose has already begun stocking her simple two-room cabin with canned provisions for the coming winter, when heavy snows often leave her trapped here for weeks at a time. She raises chickens, hogs, and rabbits, and, based on the drying pelt I notice hanging from a nearby tree, she hunts for other wildlife as well. She also has a few vegetable plots, growing just enough to have a stall at a nearby farmers' market during the summer months.

The inside of Ann's cabin is predictably austere. She's the perfect reductivist; **she has only what she needs, and uses only what she can replace. This too is a core principle of sustainability.**

I could go on. America is a big country with no shortage of sustainable stories to tell. This book has presented the stories of people who have examined how they live and decided to improve their relationship with the environment. They realize that sustainability is not a goal. It isn't a checklist with boxes to tick off. Nor is it a summit one reaches, or even a plateau. **Instead, sustainability is an ideal. Its principles reinforce that everything you do has consequences, even when they're hidden from view. Its worldview imbues you, as either a consumer or producer, with a clear, undeniable sense of responsibility.** It gives you a role—to change how you think, what you make, and what you buy—and confirms that your actions matter. Finally, it reinforces that we're both the cause of all that ails us as a society, and the solution. Only you can decide which path to follow.

"After Katrina we lost everything.
When I say 'we lost everything,' I mean
when we got out of the trees
all we had on were our pants and a smile."

SUSTAINABILITY

Any species (including Gulf Coast Fishermen and Louisiana Blue Crabs) that withstands
the impact of all user groups upon it while maintaining equilibrium throughout its life cycle

blue crabs

garfish
(used as bait
to catch crabs)

PETE

Bayou Sauvage
(tributary of Lake Pontchartrain)
New Orleans, LA
23 June 2011

WHEN I ASK PETER GERICA HOW HIS CRAB BUSINESS SURVIVED BOTH HURRICANE KATRINA AND THE BP OIL SPILL, HE SAYS, "YOU HAVE TO JUST KEEP MOVING FORWARD.

AFTERWORD
THE END (OR THE BEGINNING)

Every book has an objective. This book is no different. Every photograph, interview, statistic, and anecdote on the preceding pages has been designed with a single purpose: to explain **CLIMATE CHANGE**.

I haven't stated the term much in this book because I didn't want to scare you off, but trust me, it's on every page.

Historically, the climate change conversation has been defined by gloom and doom. Apocalyptic end-of-days stuff. The work of crisis cults. We all know what happened when people first spoke about climate change. They showed us pictures of beleaguered polar bears floating on desolate ice floes. They gave us terms like **CARBON CREDITS** and **CARBON DEBT**. We didn't understand what those ideas meant when we first heard them, and were still confused the twentieth time they popped up on our television or in a news article or at a dinner party. Continually hearing things we didn't understand didn't make them *more* clear; instead we blocked them from our minds. It's only natural. What we can't comprehend, we avoid. We tune out. Call it **CLIMATE FATIGUE**.

My interest is to take another path. If people know what terms mean, if they can see complex principles rendered simply, in ways that apply to their own lives, if they can visualize not only a complex idea but its solution, then a transformative conversation about climate change will follow.

This book has looked at sustainable approaches to food and farming. The solutions it proposes—namely building more local, transparent, and accountable food systems—are a start. Similar work needs to be done regarding water and energy, the other contributing factors to climate change. They're interconnected. One feeds off and supports the other.

The pathway to these solutions is fairly simple: Words illuminate.

They light your torch, allowing you to bring light to darkness. They are also the building blocks for ideas. Ideas form the basis of conversation. Conversation leads to consensus. Consensus creates change.

Your words can change the world.

ACKNOWLEDGMENTS

The information artworks in this book were created by the Lexicon of Sustainability project, an organization founded with my wife, Laura Howard-Gayeton. Under her direction, the Lexicon transforms contributions from literally hundreds of people around the world into a variety of sustainability initiatives ranging from films to pop-up shows, street art to educational programs, all from a barn on our farm near Petaluma, California. This book, and the larger project, would not be possible without her all-encompassing vision. I love working with you, Laura.

Our Lexicon team features an amazing collection of talented individuals, including Pier Giorgio Provenzano, Dane Pollok, Marina Veselinovic, Nicole Desanto, Samantha Harmon, Mary Tesch, Anne Digges, Ra Sol, pop-up show curators across the country, and an army of interns from Sonoma State University and Santa Rosa Community College.

I have much respect for Jessica Jones, Anna Smith Clark, Gibson Thomas, Brad Knop, Michael Eisenberg, Steven Brown, CDS, Rocky Rohwedder, and Dan Watts. I don't think this project would've happened if each hadn't lent their encouragement and support during this project's critical early stages.

I feel equally indebted to Anthony Rodale, Ken Weber, John Knox, Chis Spain, Lynne Hirshfield, Benzi Ronen, Laura Michalchyshyn, Nathan Shedroff, and Eve Saltman for their timely advice at critical junctures in the project.

Cathy Fischer has tirelessly championed our short film series, and has helped us craft partnerships with ITVS, CPB, and KQED. Thanks for believing in us, Cathy.

My thanks also extend to Aram Shumavon and Sarah Newman for mapping out the Lexicon of Sustainability's foray into Water and Energy as we continue to capture bold new solutions to climate change.

I've been fortunate to receive ongoing guidance from a knowledgeable and passionate cadre of Local Food Movement advocates in my community. They include Annabelle Lenderink, Caleb Zigas, Warren Weber, Alexis Koefoed, David Evans, Tara Smith, Jessica Prentice, Penny Livingston Stark, Tom Bensel, Brock Dolman, Kevin Lunny, Alejo Kraus-Polk, Congressman Jared Huffman, Judith Redmond, Iso Rabins, Sue Conley, Twilight Greenaway, John Lagier, Jessica Lundberg, Nikhil Aurora, Novella Carpenter, Sam Mogannam, Kenneth Rochford, Albert Straus, Paul Johnson, Nicolette and Bill Niman, Ames Morison, Richard Heinberg, Scott Davidson, Ryan Allen-Parrot, Whendee Silver, Eli Zigas, Tod Brilliant, Gloria and Stephen Decater, Doug Mosel, Bob Klein, Peter Buckley and Matt Taylor, Petaluma Bounty, Michelle Branch, Ted Fuller, and my local bakers, the inimitable Weber Family from Petaluma's Della Fattoria. Thanks to all of you for enriching the lives of so many Bay Area residents.

One thing I've learned with this project is that the Local Food Movement has no center. There are amazing people everywhere. I'm so glad to have met Mike Todd and the students of Iowa's Ames High School, David and Shannon Negus, Jon Feldman, Josiah Hunt, Mud Baron, Tyler Gray, Running Squirrel, Tony Carr, Mark Beam, Mel Weiss, David Bauer, Blu Peetz, Rachel Luster, Ian Snider, Olivia Sargeant and Jason Mann, Edwin Marty, Erika Allen, Sandor Katz, Ted Wycall, Ben Flanner, Anne Cure, Kasey White, Craig Ruggless and Gary Jackemuk, Mike and Heidi Peroni, Andrew Stout and Jessica Kagele, Hunter Lovins, Francis Lam, Jason Miel and John Gilles, Kevin Farnham, Brett Wayn, Josh Koppel,

Corky Luster, John "Farmer John" Peterson, Robert Kenner, David Burns and Austin Young, Stephen Jones, Bill Hodge, and Jack Algiere, who taught me that a carrot is not just a carrot.

A stellar collection of insightful thinkers helped shape the information artworks in this project, either by collaborating to define terms or by suggesting subjects to explore. They include Wayne Roberts, Fred Kirschenmann, Raj Patel, Brett Burmeister, Dr. Michael Hirshfield, Shannon Spring, Mae Boeve, Michael Sligh, Eliot Coleman, Jere Gettle, Bill Mollison, Lynn Henning, David Beckman, Jennifer Dianto Kimmerly, Sarah Weiner, George and Eiko Vojkovich, Colin McCrate, Julie Tilt, Eric McClam, Ann Rose, Peter Gleick, Ben Myers, Rebecca Spector, Asher Miller, Carl Safina, Jennifer Lapidus, Johannes Lehmann, John Bloom, Miguel Villareal, Adele Douglass, Michael Ableman, Danielle Nierenberg, Richard McCarthy, Frank Stitt III, George Siemon, Harry McCormack and Lynn Coody, Glenn Roberts, Deborah Madison, Patrick Holden, Will Allen, Megan Westgate, Jay Martin, John Fagen, Dr. Elaine Ingham, Gordon Woods, Mary Seton Corboy, Temple Grandin, Harriet Lamb, Geoff Lawton, John T. Edge, Sandi Kronick, K. Rashid Nuri, Thomas Kraft and Scotty Fraser, Vandana Shiva, Wes Jackson, Severine von Tscharner Fleming, Mark Kastel, Andrew Gunther, Barton Seaver, Brian Keogh, Eric Holt-Gimenez, Doug Gurion-Sherman, Philip Ackerman-Leist, and, lastly, Walter Clark and Johnny Burleson, who shared blueberries with me one fine day in early July.

We also support a number of amazing organizations. Our favorites include Laloo's, Post Carbon Institute, Zynga Foundation, Biomimicry Institute, Sustainable Food Trust, Navdanya, Polyface Farms, Culinary Institute of America, Oregon Tilth, Union of Concerned Scientists, Environmental Working Group, Full Belly Farm, La Cocina, Google Culinary Group, Southern Foodways Alliance, Alfalfa's Market of Boulder, the Brower Center, 350.org, the Land Institute, Blue Ocean Institute, Bon Appetit, Three Stone Hearth, Alba, Stone Barns, Greenhorns, Full Circle Farms, Method, Non-GMO Project, PieLab, The Organic Center, Norpac, Farmigo, Center for Food Safety, Marine Stewardship Council, Food & Water Watch, Mikuni Wild Harvest, IATP, Take Part and Participant Media, Palantir, Grist, Food Tank, EcoTrust, Occidental Arts and Ecology Center, Compton Foundation, Greenpeace, Oceana, Future of Fish, Clif Bar Family Foundation, Newman Farm, Pacific Institute, USA Artists, Grace Communications Foundation, and Pisces Foundation.

Harper Design has been a tremendous supporter of our initiative. Many thanks to Elizabeth Sullivan for her unwavering faith in the project, to Tricia Levi, who patiently fine-tuned these words, to Marta Schooler, who does fabulous work at Harper Design, and, lastly, the team of Lynne Yeamans, Tanya Ross-Hughes, Susan Kosko, Liz Esman, and Katie O'Callaghan for their invaluable contributions.

And, finally, a special note of gratitude to the 11th Hour Project. It has been a vital supporter of our project since its inception. Many thanks to Wendy Schmidt, Amy Rao, Joe Sciortino, Sarah Bell, Kevin Boyer, and Michael Roberts. Your commitment to our project means everything.

All Lexicon information artworks printed by Damian Taylor and Sanjay Sakhuja.

Credit: © Laura Howard-Gayeton

DOUGLAS GAYETON is an award-winning American multimedia artist, filmmaker, writer, and photographer. He and his wife are the cofounders of the Lexicon of Sustainability and Project Localize, which show people how to live more sustainably. He is the author of *Slow: Life in a Tuscan Town* and lives on a farm with his wife and daughter in Petaluma, California.

www.lexiconofsustainability.com
www.facebook.com/thelexicon
twitter.com/lexiconproject

HarperCollins books may be purchased for educational, business, or sales promotional use. For information, please write: Special Markets Department, HarperCollinsPublishers, 195 Broadway, New York, NY 10006.

First published in 2014 by:
Harper Design
An Imprint of HarperCollins*Publishers*
195 Broadway
New York, NY 10006
Tel: (212) 207-7000
Fax: (212) 207-7654
harperdesign@harpercollins.com
www.harpercollins.com

Distributed throughout the world by:
HarperCollinsPublishers
195 Broadway
New York, NY 10006
Fax: (212) 207-7654

Design by Tanya Ross-Hughes, Hotfoot Studio
Photographs on page 233 (top right and middle left) by Kait McKinney

Library of Congress Cataloging-in-Publication Data: 2013951754
ISBN: 978-0-06-226763-4

Printed in the United States